This book examines industrial relations in the lithographic industry, explains its separation from the industrial relations system in other branches of the printing trade and shows the development of present collective bargaining practices.

Divided into three parts, the study first sketches the economic and technical setting of the industrial relationship, then describes the factions and their past relationships; the final section examines the current bargaining relationship and the wider consequences of bargaining.

In his foreword, Dr. Charles A. Myers, Professor of Industrial Relations at M.I.T., says, "The value of a study of this sort lies not only in the careful analysis of all facets of the labor relations experience in historical perspective in the lithographic industry, but in its broader significance for the study of labor relations generally. What Munson does is to show, in a wealth of detail and illustrations, that the work rules developed by this industry's industrial relations system are a consequence of the interaction of the 'actors' (labor, management, and government) with the changing technology and market structure of the *lithographic* industry. This sort of research is an antidote to easy generalizations about collective bargaining, management prerogatives, employer associations, seniority, job control, and union attitudes toward technological change, to mention only a few."

The author points out that the effect of conventional market factors, such as insulation from competing products, closed labor markets, and so forth, on wage rates is clearly seen. But of more interest is the effect of these and less obvious factors on employer bargaining strategy, on the structure of employer bargaining units, on union support of flexible wage setting devices, and on the internal structure and decision making processes of unions.

This volume is the sixth in a series on labor-management history.

Fred C. Munson teaches industrial relations at the University of Michigan.

Labor Relations
in the Lithographic Industry

𝕾 𝕾 𝕾 𝕾 𝕾 𝕾 𝕾 𝕾 𝕾 𝕾 𝕾 𝕾 𝕾 𝕾

FRED C. MUNSON

HARVARD UNIVERSITY PRESS

Cambridge, Massachusetts

1 9 6 3

To
J. PAUL MUNSON
and
JOH H. MUNSON,
to whom the six,
and now the forty-two of us
owe very much.

Foreword

The concept of an industrial relations system as a useful method of analysis has been applied with precision in this book to the study of labor relations in a small but significant industry. While the printing and publishing industry as a whole is large, employing about 865,000 in 1958, less than one eighth of these were part of the lithographic (offset printing) work force. The union represents about 38,000 production workers. Few thorough recent studies have been made of collective bargaining in the various branches of the printing industry, and this book is a welcome addition to the Wertheim series on collective bargaining in particular industries, under the general editorial direction of John T. Dunlop.

The value of a study of this sort lies not only in the careful analysis of all facets of the labor relations experience in historical perspective in the lithographic industry, but in its broader significance for the study of labor relations generally. What Munson does is to show, in a wealth of detail and illustrations, that the work rules developed by this industry's industrial relations system are a consequence of the interaction of the "actors" (labor, management, and government) with the changing technology and market structure of the *lithographic* industry. This sort of research is an antidote to easy generalizations about collective bargaining, management prerogatives, employer associations, seniority, job control, and union attitudes toward technological change, to mention only a few.

Collective bargaining is old in the lithographic industry, and Munson points out that the earliest records go back to the 1850's. A national employer's association was established late in the nineteenth century to counteract the growing power of the union, and national collective bargaining was established during 1904–1906 and again for a short time after 1918. Except for a brief period during the NRA, it later gave way to increasing local employer association bargaining. Munson shows how changing technology, particularly innovations in lithography, affected the market structure of the industry so much that employers became less homogeneous and had less unity at the bargaining table as a consequence.

These same factors progressively reduced the average size of lithographic firms, many of which are still family-dominated and/-oriented. This has significance for a consideration of managerial prerogatives. Few management officials of these firms perform in fact the managerial functions of recruitment, selection, wage administration, or union contract negotiation. They have been forced to delegate much of this to professional spokesmen in local employers' associations, and, despite legal fiction, the union is still the effective source of skilled men. The extent to which our generalizations about managerial functions in personnel administration, growing largely out of experience of larger firms in major industries, have to be modified for industries with market structures like that in lithography, deserves more research. Munson has pointed up briefly the digressions from the model in this industry.

If changing technology and market structure have created problems for employers, they have also affected the union, which now faces a crisis. The steady erosion of market boundaries by the extension of lithographic processes into plants (such as can factories, paper products firms, and so forth), where there were already other bargaining units and representatives, has led the union into jurisdictional conflicts with other unions. Here the role of the National Labor Relations Board has been crucial, and the success of the lithographers' union in establishing its right to represent the "traditional" lithographic unit within a larger bargaining unit has enhanced the role of the general counsel of the union to a point far beyond the typical position of a "staff" man in the union hierarchy. Again, Munson has forced us to re-think traditional concepts about the power structure of unions. Similarly, since technological change in lithography has *expanded* markets and the number of jobs, it is not surprising that the union actively supports improved technology; but this is contrary to the popular belief about union attitudes.

The union's position that "we are unable to know in advance into what other industry lithography will penetrate" has led it to reject the "no-raiding" pact and withdraw from the AFL-CIO in 1958. It has powerful union adversaries in its practice of "reorganizing the organized," and Munson predicts that further technological and market changes will, in fact, reduce the distinctions between processes, and force the union into efforts to get working agreements with rival printing unions, or, alternatively, support from employers. In the face of coming innovations, neither alternative looks too promising. Munson's rather gloomy prognosis is: "It is entirely possible that interunion conflict will have to become far more acute than at present before there will be sufficient member interest to permit amalgamation." In the meantime, many lithographic employers may be caught in the middle.

The work rules which have developed out of the system of industrial relations in this industry are complex, and Munson's discussion deserves careful study. His generalizations in the final chapter, on the circumstances and conditions under which work rules can change, are especially pertinent in the current discussion of what employers call "restrictive" work rules and unions term "protective" rules. In particular, his discussion of some considerations when modifying the rules of an industrial relations system should be required reading for all proponents of legislative remedies to the work rule problem, as well as for students of industrial relations systems.

Finally, I want to add that this study was begun as a doctoral thesis in the Industrial Relations Section of the Department of Economics and Social Science at M.I.T., under the supervision of Douglass V. Brown and myself. But the present book is the result of further study and reflection by the author, as any good doctoral thesis should be when it merits publication.

December 1962   Charles A. Myers
              Massachusetts Institute of Technology

Contents

Tables

Illustrations

Author's Preface

One of the pleasant parts of an author's job is to acknowledge the help of people who have made his work possible. In a study involving substantial field research there is little hope of giving all who helped the credit they deserve, and in the following paragraphs I have been forced to single out those people whose assistance has been substantial.

Among employers, three people in particular have given unsparingly of their time. Kenneth Scheid, formerly with Forbes Lithograph of Boston and now a Professor at Carnegie Institute of Technology, introduced me to the industry, and has assisted and advised me throughout the two years of research. George Strebel of the Printing Industries Association of Western New York clarified the pattern of changes I should be alert for, and has answered many calls for assistance. William Winship, formerly of Brett Lithograph in New York, corrected several misunderstandings I was harboring, and together with the two men above, spent much time in providing helpful criticism of the manuscript.

I have also received an unusual degree of cooperation from the Amalgamated Lithographers of America. Past local President Arthur Willis of ALA Local 3 in Boston patiently endured many hours of questioning, and help was also received from the ALA local officers in Rochester, Buffalo, Toronto, New York and Baltimore. Local 4 President Harry Sphonholtz in Chicago was particularly helpful. The international officers of the ALA, notably Secretary-Treasurer Donald Stone, President Kenneth Brown, Vice-President Jack Wallace, former General Counsel Benjamin Robinson and past Vice-President Martin Grayson, now Regional Manager of Printing Developments, Inc., of Chicago, have allowed me not only full use of historical records, but the transfer of many documents to Cambridge and later to Ann Arbor where I could dig in them at my leisure. Donald Stone and Martin Grayson have been particularly useful in giving me an understanding of the union. Much of the historical information about the Amalgamated Lithographers of America gathered during the research appears in the author's *History of the Lithographers' Union* and is not repeated here.

Howard Patterson of the Graphic Arts Institute of New England, together with Douglas Reilly, Arthur Howard and Kenneth Scheid of the Boston employers' group, permitted me to attend the 1958 Boston negotiations as a guest of the employers. The degree of local cooperation is indicated by my further admission to the employer strategy sessions during the negotiations.

Other employers, among them Carl Reed of Buffalo, James Armitage and William Reinhardt of Chicago, and Winslow Parker and Bennet Young of Baltimore, have provided useful material. Trade association executives George Mattson, Maurice Saunders, Oscar Whitehouse and Arch McReady have supplied much factual data and many comments.

Without the quite remarkable cooperation I have received from these employers, trade association executives and union officials, this study could not have been made. Their cooperation stemmed from a desire to help in anything that might benefit the lithographic industry, and I hope they will not feel their time was wasted. To all of them I express my thanks.

Professor John T. Dunlop of Harvard University, Professors Dallas Jones and George Odiorne of the University of Michigan, provided guidance, as has Professor Charles A. Myers of M.I.T. The assistance received from Professor Myers goes well beyond the confines of this book, for I had the privilege of studying under him at M.I.T. He and Professor Dunlop gave invaluable advice in the early stages of this study, and I have drawn heavily on Professor Dunlop's ideas in organizing this work. Miss Grace Locke, and Mrs. Raymond E. Crabtree have been most helpful.

Initial research for this study began in 1957, and the text was prepared in 1958 and 1959. Some revisions, primarily in census data, were made in 1960. Much of the work was done under a grant from the Social Science Research Council. The financial support provided for the necessary field research and food and clothing for my family. Any one who has been a married graduate student will understand my warm affection for the council.

July 1962   Fred Munson

Labor Relations
in the Lithographic Industry

Abbreviations

ALA	Amalgamated Lithographers of America
CPLA	Conference for Progressive Labor Action
IALS & PPA	International Association of Lithographic Stone and Plate Preparers
ILPA	International Lithographic Protective Association
IPEU	International Photo-Engravers' Union
IPLA & PF	International Protective Association of Lithographic Apprentices and Press Feeders
IPP & AU	International Printing Pressmen and Assistants' Union
ITU	International Typographical Union
JLAC	Joint Lithographic Advisory Council
LIP & BA	Lithographers' International Protective and Beneficial Association
LNA	Lithographers' National Association
LTF	Lithographic Technical Foundation
NAEL	National Association of Employing Lithographers
NAPL	National Association of Photo-Lithographers
NLRB	National Labor Relations Board
NRA	National Recovery Act
PAA	Poster Artists' Association of America
PIA	Printing Industry of America
USP&L	United States Printing and Lithograph Co.

Chapter I

The System

Labor problems in lithography are as old as the process itself. Alois Senefelder, the inventor of lithography, was the victim of the first in 1796, when he hired help to meet a fast approaching deadline. "Oh human weakness!" Senefelder lamented. "Does it seem credible that of my six helpers not one could master the extremely simple method, the mere matter of rubbing evenly and thoroughly?" [1]

Occasionally one finds employers who still accept Senefelder's conclusion that labor problems come from human weakness in the other fellow. But today the lithographic (offset printing) industry has increased its employment from six workers to 100,000, and the analysis of labor problems requires a broader base than the study of human frailties. There are, after all, real difficulties involved in deciding how much a worker shall be paid, how long he will work each week, what he will and will not do, how these matters will be decided, and who will represent him in deciding them. These questions are often defined in an economic framework and treated as one set of problems which arise within the economic system. But an economic framework works best with economic problems, and labor problems are more than that. They take shape in a context which is both complex and dynamic. In lithography, changes in technology have led to changes in the structure of the industry, as well as to changes in the product market for lithography and the skills required in the lithographic workforce. Each of these has had its effect on working rules and the rule-making relationship. Public attitudes about labor problems have been a significant part of the context, notably when they have been given form and substance through legislation. And, always, a part of the context has been the body of rules and practices already developed — history itself.

The oldest and still dominant union in the industry, the Amalgamated Lithographers of America, traces its parentage to a union founded eighty years ago. And a still-vigorous jurisdictional dispute with other printing unions has forty years of history behind it. Collective bargaining itself has a venerable history with national contracts having been introduced shortly after the turn of the century. An analytical framework is clearly required to weigh the importance of a new process, a cherished tradition, or a change in the law, and to trace its effect in the rule-making process. The one adopted draws heavily upon that proposed by John T. Dunlop, which is described and its usefulness illustrated in his recent book, *Industrial Relations Systems*.[2] The bare outlines of this framework will be provided here.

The Concept of an Industrial Relations System

Professor Dunlop sees the industrial relations system as a subsystem of the larger society. In much the same way that the purpose of the economic subsystem is to produce goods and services, the purpose of the industrial relations system is to "produce" the rules of work which are acceptable to both parties, and to provide acceptable methods for determining what these rules shall be. In the economic system men determine what they shall consume; in the industrial relations system they determine under what conditions they will work to produce it. This network of rules, which may be called the "output" for the industrial relations system, concerns all aspects of the employer-worker relationship. Wages cannot be separated from the total web of rules for they are determined in the same context by the same rule-makers and for the same set of purposes as all other rules.

The rule-makers in the industrial relations system are the workers and managers and their representatives, and specialized governmental or non-partisan private agencies concerned with the worker-management relationship. They may and often do have a common set of ideas and beliefs that help bind the system together. This ideology shapes the relationship but has little effect on the content of agreements between unions and managements, or decisions by arbitrators or government agencies. Yet the rules are far from being fashioned in a vacuum. The rule-makers are subject to a bewildering variety of pressures, which nevertheless can be dealt with usefully if one sees them as stemming from three basic sources: technology, market pressures, and the distribution of power in the larger society.

(1) *The technological characteristics of the work and the workplace.* One may expect different rules (and different organizations among the rule-makers) if the workplace is isolated — as in coal mining, or is a roving site — as in construction; if the work is machine-paced — as on an assembly line, or highly-skilled — as in the case of patternmaking. To these static considerations must be added dynamic ones, such as when advancing technology decrees a change in the optimum size firm, a redistribution of work

processes among firms, or an obliteration of formerly distinct craft lines.

(2) *The market pressures.* It is obvious that the price a lithographer receives for his goods will affect the wages he can pay his employees, and that this price is determined in competition with other sellers of lithography. This market for lithographed goods is not only important in effecting the level of wages. The nature of the market plays a role in determining the size of the bargaining unit (single-firm, city-wide, or national), the seriousness of an interruption of production, what demands will be placed on employees (in their relations with customers, for example), and so forth. Indeed, the classic approach to the study of the employer-employee relationship assumes that all important pressures, including technology, operate through the market place, and it is suggestive of the bias of most scholars that this field of study is called "labor economics."

Such an approach claims too much for the market. Market pressures are only part of the total context in which an industrial relations system operates, albeit an important part.

(3) *The locus and distribution of power in the larger society.* The importance of the first two basic pressures — the technological characteristics of the work and workplace and the market pressures — is self-evident. This third element, the locus and distribution of power in the larger society, is more inclusive than the other two and can be more easily illustrated than described. For example, the decade of the 1920's in which the Supreme Court fashioned a law against unions from a law (the Clayton Act) which exempted them can be readily distinguished from the decade beginning in 1935 when the National Labor Relations Board was entrusted with the sole function of protecting and encouraging the unions' growth. To take another example, a trade union movement which believes itself to be at odds with the larger society will develop class consciousness, while one which is accepted by society may even cease to be a movement and be better described as "the union business." Again, in cases of conflict between worker organizations and management, the acceptable weapons of combat (boycotts, industrial spies, strikes, blacklists, yellow-dog contracts) as well as the terms on which peace can be made, will depend on the influence of the actors in the larger society. This influence, or status, will shape the formal rules of the game as enforced by government, and the informal rules as enforced by public opinion. Such status will also affect the union's self-image and will help determine whether the union leadership is radical, conservative, idealistic, or businesslike. One cannot explain the ritualistic secrecy of the Knights of

Labor, the fast-fading socialism of the early AFL leaders, or the rise and decline of the IWW without reference to the attitude toward unionism in America, and the effect of this attitude on the unions themselves.

This threefold division of the context in which the rules of the workplace are forged is a framework within which facts may be organized. It must not be misunderstood as a law of nature or society, which can be adjudged true or false. It is logically more akin to a filing system, useful in organizing the great variety of information relevant to a study of industrial relations. Since the purpose of this book is to show not only how the working rules in lithography are defined by the actors, but to show how these definitions are in large part determined by the context, such a framework has obvious value.

The Lithographic Industry[3]

Lithography or offset printing is the name given to one of the three important methods of printing. The oldest and still most commonly used is printing from a raised image; this is called letterpress or relief printing. A second method is gravure, which uses a depressed image rather than a raised one. The gravure process has not found wide application outside the longrun publications and packaging field, and in this study it is convenient to include it in the letterpress grouping. Lithography uses an image which is on a plane with the rest of the plate, hence the occasional use of the phrase, "planographic printing."

The lithographic industry, as the term is used here, includes firms using the lithographic process which compete with each other in hiring workers with lithographic skills or in supplying customers with lithographed products. Open and vigorous competition is not evident either in hiring or selling, and one could describe the labor market connection as one linking all firms which can offer a lithographic craftsman work in his trade, and the product market connection as one linking all firms that are equipped and willing to bid on an order for lithographed products.

This definition is somewhat broader than that used by the Bureau of the Census, which identifies all firms using the lithographic process but includes them in "Standard Industrial Classification (S.I.C.) 2761 Lithographing" only if the greatest proportion of their output is in lithographed paper products. Nevertheless, the Census definition includes the greatest part of the industry for our purposes, and much of the detailed information provided for this group or firms is relevant to the industry as a whole.

Measures of size. Interest in the lithographic industry is justified by its recent and quite remarkable growth. In 1899 it had 15,000 employees, in 1935 scarcely 5000 more. Twelve years later in 1947 employment had increased to 52,400 and by 1958 to 97,500. By contrast, the letterpress industry has grown at a much slower rate in the present period, its present employment of 200,000 being less than a 50 per cent increase over 1939 and scarcely a 4 per cent increase over 1947. While lithographic firms account for only 27 per cent of all commercial printing sales currently, their relative importance has steadily increased for the past twenty-five years, and seems likely to continue doing so in the future. This trend is readily apparent in Table 1.

Table 1. Growth of lithography and printing, 1939–1958.

Year	All employees (thousands)				Value added (millions of dollars)			
	Lithography	Letter-press	All commercial printing[a]	Col. 1 as a per cent of Col. 3	Lithography	Letter-press	All commercial printing[a]	Col. 5 as a per cent of Col. 7
1939	34.7[b]	129.1[b]	229.3[b]	15	99.9	381.9	664	15
1947	52.4	191.7	329.7	16	314.1	970.3	1725	18
1954	77.7	200.2	367.2	21	578.9	1361.8	2509	23
1958	97.5	199.3	391.8	27	872.4	1591.0	3203	27

Source: *1958 Census of Manufactures*, vol. II, pp. 27B–4, 27B–5, 27C–4, 27C–5.
[a] "All commercial printing" refers to the following census industries:

	Value added (million dollars) 1958
S.I.C. 275 Lithographing	872
S.I.C. 276 Commercial printing	1591
S.I.C. 278 Bookbinding and related industries	286
S.I.C. 279 Printingtrade services (typesetting, photo-engraving, etc.)	454
	3203

In order to maintain comparability with earlier data, 1958 figures have been adjusted to nclude establishments now classified in the manifold business forms industry.
[b] Estimates; raw data from *1939* and *1947 Census of Manufactures*.

A unique aspect of the lithographic industry's growth has been the accompanying decrease in the average size establishment. For example, shops employing less than twenty people made up half the establishments in the industry in 1929 and three quarters in 1958. To express the same development somewhat differently, establishments that were a third or less the average size, measured by employment, made up 35 per cent of the industry in 1929, over 50 per cent in 1958. The "average" establishment itself has some significance; in 1929 it employed sixty-three people, in 1958 only twenty-four. From 1947 to 1958 alone, total employees in the industry increased 86 per cent, while the number of establishments increased 160 per cent. Most of the small shops serve local markets, and so are spread widely throughout the country. In 1921 there were nine states with one lithographic establishment and twenty-one with none at all. In 1958 only North and South Dakota, Nevada and Wyoming had less than five establishments.

Patterns of integration. One qualification should be noted in regard to the declining average size of establishment. Census data are for employing units, several of which may be in a single firm. Actually, firms having two or more establishments account for 40 per cent of total employment but less than 2 per cent of total establishments. The outstanding example of horizontal integration is provided by United States Printing and Lithograph Co. In 1891 United States Printing was formed by a merger of five companies, the major one (Russell and Morgan) itself a result of an earlier combination. Ten years later in a quite separate venture United States Lithograph was formed and in 1905 became Consolidated Lithograph, combining seven lithographic firms with an estimated 90 per cent of the poster business in the United States. United States Printing and United States Litho (Consolidated Lithograph) were brought under one management in 1912 and formally merged in 1915, forming the United States Printing and Lithograph Co. (USP&L). USP&L added seven other major firms between then and 1929, among them the American Lithographic Co. — a combination of ten separate firms formed in 1892. Since 1929, USP&L has added eight large lithographic firms,[4] so that in 1959 its ancestry could be traced to about thirty-five individual firms, although it operated only nine separate plants.

The USP&L story vividly illustrates that mergers have taken place, but it also illustrates that they were taking place fifty years ago. In point of fact, the degree of concentration in the lithographic industry may well be less today than at any time in the past. The combination of existing firms always attracts more attention than the birth of new ones, and each generation tends to think the curse of bigness is descending on it with a special vengeance. Certainly Census data do not show such a trend in lithography. Yet there is the possibility that an important change is occurring, as large companies, particularly in the paper industry, buy up existing large lithographic firms. Numerous examples of this are available, one of the most notable being the recent acquisition of the same USP&L described above by Diamond Gardner Corporation, one of the major paper companies. It may well be that in this segment of the lithographic product market (lithographing packages and other containers), "the current wave of mergers" is resulting in a distinct increase in the average size of the employing firm.

Vertical integration *by* the lithographer, rather than *of* the lithographer, is less common. Few lithographers have ever built their own paper plant, and it is clear that establishments of the 1920's were more integrated than most present-day firms. There is an increasing tendency to purchase premixed inks, premixed chemicals, presensitized plates, all of them operations formerly carried on within the shop. The first offset press for paper lithography was in fact made by a lithographer, Ira Rubel in New Jersey.[5] Today even the dampening-roller covers are purchased readymade.

There are few examples of integrating into publishing or advertising, and efficiency seems best served by two managements cooperating rather than a single management coordinating. The problem is twofold. Lithography is fundamentally an end product; the buying firm may use it to help its customer's customer's customer sell more dog food, but with few exceptions the final operation on the product has been performed when it leaves the lithographic firm. No technological economies are inherent in attaching a printing plant to General Motors or McGraw-Hill, although other benefits may be gained. Second, a lithographer integrating forward would in most cases be like an ant trying to swallow an elephant. There are few buyers of printing for whom it is more than an exceedingly small portion of their total purchases. The real question of integration arises when a very large firm decides to run its own printing plant, or when technological economies can be secured by integrating the lithographic operation into a production process, as for example in the imprinting of tin cans to be used for containers.

A further development reducing the degree of vertical integration is the growth of the trade shop. This type of firm, common in the letterpress industry, specializes in a single function such as typesetting, platemaking (electrotypes or photo-engravings), or bindery work. Trade shops are by no means so common in the lithographic industry as they are in commercial printing, and there are still large cities which make almost no use of them. Where there are only a few lithographers in a city, a trade platemaker leads a precarious existence, for his function insofar as a large shop is concerned is to handle overloads in the preparatory department. But the small lithographic firm is small in part because it does farm out all of certain operations and thus gives the trade shop a somewhat less volatile market. Additionally, as lithographic firms move into book printing and other products requiring extensive typesetting, they turn to the trade shops for their type, and to the trade binderies for the required finishing operations. Trade shops are becoming an integral part of the lithographic industry, tying lithographers ever more closely to the printing industry.

Other Users of the Lithographic Process

The foregoing section has described the characteristics of exactly 4259 establishments only because the Bureau of the Census has found exactly that number which meet its necessarily precise definitions. Thus, two partners who run a small firm without hiring additional labor will be excluded from the Census-defined industry, but an owner with one employee will be counted. A lithographic establishment, large or small, may be included in the steel, insurance, auto, paper products, or other industry if its serves only the internal needs of the firm which owns it. Finally, a firm with 50.1 per cent of its sales volume lithographically produced and the balance letterpress will be included in the lithographic industry, but if the proportions are reversed the firm "changes industries" by Census definition. Because of this latter definitional practice, 21 per cent of all lithographed output is by firms which are in other industries, primarily letterpress. Yet even this phrase "lithographed products" has a Census definition considerably less inclusive than all lithographed production. Lithographed books, greeting cards, cardboard boxes, periodicals, newspapers, and most lithographed metal products — in fact nearly all lithography except commercial printing — are excluded from the definition. Were both the skills of the workforce and the market of such firms entirely different from those of lithographic establishments, they would hold no interest for us. Such is not the case, and it is accordingly important to describe their characteristics.

No-employee firms. Perhaps the least important of the firms excluded from the Census definition are those with no hired employees. Of the producing units among this group the largest proportion of sales is to other printers rather than to final users. A number of other "firms" are printing brokers who perform only the selling function and buy their printing from producer firms.

Commonly heard estimates of the number of such units in existence range from 2000 to 10,000. There are at least fifty listings under "lithographers" in the Boston Yellow Pages unfamiliar to a man of twenty years' experience in the Boston graphic arts industry. If this were the number of no-employee lithographic "firms" in the Boston Metropolitan area, it would imply that there were some 2500 in the country at large.[6] The one point of some significance is the evidence that at least, in a broad sense, both capital and labor inputs are almost infinitely divisible.

Captive plants. "Captive plant" is the name given to productive units which service only the firm which owns them. In general, three types may be distinguished.

Type I: General Motors, the United States government, and many other large and small users of printing operate lithographic equipment to produce internally consumed forms, drawings, bulletins, and so on. Some of the largest may also use their captive plant operation to produce externally used items such as advertising material, insurance policies, securities, Census bulletins, and so forth. These "private plants" may be no more than a single offset duplicator running solely from presensitized-paper plates "made" by placing them in a typewriter and cutting them as one would a stencil. There are at least a half-dozen such "private plants" in every major university. At the other end of the scale, there are private plants owned by large users of lithographed materials, which may or may not compete actively for outside business in addition to meeting the owning firm's requirements for printing, but which, in any event, operate as a profit center separate from the rest of the corporation.

The diseconomy of private plant operation most often mentioned in interviews with employers is the problem created by having different working conditions for parts of the same workforce. A plant of any size is fair game for the lithographic or other printing unions, and the best but not always successful way of keeping them out is to match union working conditions. Wages, hours, overtime, and other job-control provisions common to lithographic plants can provide endless grounds for unrest and dissatisfaction in the remainder of the workforce when applied to a select few working under the same roof. The unwillingness of managements in other industries to risk having the pattern already established in lithography spread to the rest of their workforce has tended to keep the number of substantial private plant operations at a minimum.

Type II: The second type of captive plant uses lithography as an integral part of the production process. For example, Continental Can Company lithographs the outside surfaces of containers, the operation being just one of a series in a semicontinuous flow from sheet steel to packaged containers. Similarly, paper mills have added lithographic presses to imprint the paper board prior to the box assembly operation. Continental Can or the Sutherland Paper Company, and all others which have lithography as an integral part of their operation, reduce the business potentially available to lithographers in the same way as does the first type of captive plant, but, much more important, they compete to supply the same customers.

In metal decorating, captive plants have been the typical organization for over four decades. An informed estimate places total lithographic employment in independent houses at not more than 500 employees.[7] The addition of lithographers employed by captive plants raises the lithographic workforce figure from 3000 to 3500. This is a relatively small number of workers, in view of the fact that metal lithographers compete with each other, but less so with paper lithographers. Developments in the food-processing and packaging industry may increase competition between metal and paper lithography, but the limits of direct competition currently are the choice of using a paper label or lithographing directly on the metal container. Partly because of this, and also because of the general tendency for metal lithographers not to take advantage of their interchangeable skills and move into paper houses, collective bargaining in the metal decorating field is somewhat separate from the bargaining in paper lithography. It is strongly affected by patterns set in the paper houses, but there is negligible pressure in the other direction. In New York City, for example, the key bargain is made between the ALA and the Metropolitan Lithographers Association. Separate contracts are signed between the ALA and the local metal decorators and captive plants, yet neither the employer nor the union spokesman interviewed suggested it was anything but a carbon copy of the key bargain in all essential respects. The employer spokesman for the independents, when asked why the metal decorators did not join the Metropolitan so they would have a say where it counted, said that they had considered joining but were so far outweighed by the paper houses that it was not worth the effort.[8]

Unlike the metal decorating plants, Type II captive units which lithograph on paper are a fairly recent development and are found primarily in the folding paper box industry. As recently as 1940 the two major manufacturers of large lithographic presses sold no equipment to producers of folding paper boxes; in 1958 perhaps 40 per cent of all large and medium-size presses were sold to them.[9] One large paper company has installed four large five-color lithographic presses since early 1955; before that it had none in operation.[10] In this type of operation there is little possibility of segregating the lithographic operation from the rest of the production process, and the prospect of having to deal with the lithographic union is so distasteful that in one case a management spokesman has asserted in NLRB hearings that the company would discontinue the operation first.[11]

While changing technology has been the significant factor in facilitating the growth of Type I captive units — those which produce internally consumed printing — the increase in lithographic units in the box-making industry comes more directly from a response to pressures from the market. An estimate of sales volume in the folding paper box industry shows the following pattern of growth:[12]

1940	$135 million
1947	$480 million
1950	$500 million
1958	$900 million

As the demand for extensively decorated cartons and containers has grown in volume, it has become profitable for producers of paper board to enter a market formerly served most effectively by large job shop operations. Nevertheless, innovations are in part responsible for permitting the lithographic process to be used in supplying this market. Formerly, any printing in the industry was done almost exclusively by letterpress. Two developments — the successful introduction of coatings for paperboard that will resist being "picked off" the base material by the tackiness of the lithographic ink, and the long-run lithographic plate that can consistently produce up to a half-million impressions even under relatively severe conditions for the lithographic method — have led the industry to introduce lithographic equipment for much of the increased volume. Competition from the paperboard industry has a potentially heavy impact on the lithographic and letterpress industries generally, and is one of the major problems facing printing unions and managements today.[13]

Type III: The third type of captive plant is the publishing firm, which lithographs part or all of its publications. In industry parlance such units are not considered captive plants, although they are indistinguishable in their economic relationship to the industry.

The important published items which may be produced by lithography are books, greeting cards, periodicals, and newspapers. Except for greeting cards, the lithographic process is a recent entrant to these product markets. Yet since 1939 lithographic employment in publishing has been growing at twice the speed of the expanding lithographic industry, and, like the folding paper box industry, publishing firms are becoming important employers of lithographic craftsmen and producers of lithographed products.

Combination plants. The final group of firms which are excluded from the industry definition are called "combination plants" — printing firms which use both the lithographic and letterpress process for printing, but whose major source of revenue is from letterpress production. Nearly all are former letterpress firms. As early as 1909 lithographic employers were worrying "that in the very near future every little crossroads printing shop will be equipped with an offset press," [14] but there were very few printers that cared to experiment with the method until the 1930's. By and large, the instability of the lithographic plate when not prepared and run by workmen thoroughly conversant with the lithographic process restricted its use to lithographic houses. In about 1932, however, the Multilith, a small semi-automatic offset press, was introduced to the trade and substantially reduced the skill requirements for operating an offset press. The problem of making plates could be given to trade shops, where they existed, thus making it possible for letterpress houses to introduce the offset process without requiring much addition to the knowledge of either the management or the workforce.

The war and immediate postwar period brought with it basic simplifications of the platemaking operations, and therefore also a rapid increase in the number of combination plants. Noneconomic factors also helped lithography grow:

> An unreasoned conviction prevailed in the minds of purchasing agents that any printing produced by offset would be cheaper. This was fostered during and after the war by articles in the trade papers . . . written largely by writers only partly informed. Consequently many jobs were specified offset which could have been produced as well or better by letterpress. Then too, when the letterpress printer, after considerable soul searching, acquired the offset press, his attitude was that of a child with a new toy. He diverted to it jobs which should have remained letterpress. In almost any shop there can be found examples of such jobs which have since been retransferred to their original process.[15]

By 1948, nearly one half of all small offset presses were being sold to letterpress houses just going into offset, only 11 per cent going to straight lithographers.[16] By 1959 one estimate placed the number of combination plants at two thirds of all printing firms (excluding bedroom shops), and the answers of the firms surveyed indicated that this proportion would rise to 72 per cent in 1960.[17]

Although Census data are of only limited value here, they do confirm the trend toward combination shops. Table 2 indicates in which industry commercial lithography was produced and shows the rapid introduction of the process in the letterpress industry after 1939. A change in industry definitions makes it impossible to show comparable 1958 figures. The apparent reversal in the proportion of commercial lithography done in the commercial printing industry in 1954 probably is due to the definitional crossover of former letterpress plants into the Census-defined lithographic industry as that process became the major source of revenue. As already noted, the Bureau of the Census assigns a firm to an industry according to its major source of revenue. A sufficient number of such crossovers occurred to cause the bureau to run a special check on them, but, regrettably, no further use was made of this material after it was gathered.[18] The number of establishments in the Commercial Printing industry increased by 10 per cent from 1947 to 1958, while in the lithographic industry the number increased 200 per cent, so the argument for a large number of crossovers is rather strong. An estimate has placed the number of plants having offset presses 17″ x 22″ or larger at 5162 in 1957, but it is difficult to relate this figure to the Census figures.[19]

Table 2. Distribution of commercial lithography by industries. (figures in millions of dollars)

Year	All industries	Lithographic industry	Commercial printing industry	Lithography as a per cent of total commercial printing industry sales volume	Per cent of lithography done in commercial printing industry
1925	108.0	93.8	11[a]	1.6	10
1929	130.6	115.5	12[a]	1.4	9
1939	157.7	123.4	18[a]	3.0	11
1947	465.5	359.2	88.9	6.0	19
1954	999.1	791.0	156.2	8.0	16

Source: *Census of Manufactures* for the years given. Basic data appear in Appendix A.
[a] Estimated by taking 75 per cent of all lithographic production produced outside the lithographic industry.

Chapter II

Key Actors in the System

Employers and their Organizations

The typical lithographic firm is too small to be an effective unit for bargaining. Its small size and low capital, which in 1958 was between $5000 and $6000 per employee, does not provide the resources to resist a long strike. Equally significant, a small firm can rarely afford to compete in the market for professionally qualified managers. In recent years there has been an increase in the proportion of unincorporated firms, nearly all of which are family-owned and operated. Even among incorporated firms a substantial number appear to fall within this category. A review of *Poor's 1958 Register* indicates that half the listed corporations were family-owned and operated.[1] In the Boston area there are not more than two or three firms where ownership and operation are not clearly centered in a maximum of three families.[2]

Of the three potential sources of management — the family, the firm, and outside recruitment — only the first two are actively used. As a result, lithographic managements are often top-heavy with men who entered with much skill in lithography and none in management. This is even true of family members.[3] The extent of this problem is highlighted by the following experience, related to the writer by the key management member of a successful family-owned and -operated firm. He had been asked to address a gathering of printers in another city, and in the question period which followed was asked what could be done about getting good management and good foremen:

> How many of you have sons? [About 25 or 30 raised their hands.] How many work with the business? [About 3 raised their hands.] . . . So I told them, Here's the reason you're getting poor management. The family isn't following in the business. You can't expect them to if you don't pay them as much as they can get driving a truck or working as a papermaker.[4]

It is reasonable to expect that in an industry where family management is a virtue, the over-all quality of management will not be high. So serious is this problem in the area of cost control for example, that accounting ratios based on reports of cooperating firms are believed to have little value for the smaller firms because the raw data is so unreliable.[5]

Summing up a discussion on the usefulness of published cost data generally, another trade association interviewee said that "in spite of the millions of dollars that have been spent on cost education, this industry is just a jumbled mess." [6]

An even more pronounced absence of managerial competence is evident in the handling of employee relationships. Relatively few firms have a management member with the ability, inclination, and, perhaps most important, the time to deal capably with the full range of personnel problems. A most significant weakness is the lack of professional competence in the field of union relations. One result is that union relations cover nearly the entire field of employee relations. Another result, both of unskilled management and the financial weakness of the small firm, is that such firms can improve their position materially by binding together in trade associations.

Only two types of trade association are of interest here, the national lithographic employer associations and local associations formed primarily for the purpose of collective bargaining.

National employer associations. The three national associations have little direct contact with collective bargaining. From 1888 to 1933 there was only the Lithographers National Association, but in the latter year the National Association of Photo-Lithographers was established to provide representation for small commercial lithographers. As the number of combination plants increased, the association that represented them when they had been letterpress firms, the Printing Industry of America (and predecessors), also took on the role of a national association for lithographers. Letterpress firms much prefer the phrase "offset printers," and this is normally used by the PIA to describe this segment of its membership.

The Lithographers National Association.[7] Predecessors of the Lithographers National Association (LNA) go back to 1888, but the introduction of a strong and unified association dates from 1906 when the National Association of Employing Lithographers was formed with the definite purpose of "settling the union's hash" which it did with dispatch and efficiency. Since that time the power of the association has declined. Today the major activities of the LNA are advertising and general promotion of the offset process, providing management aids in such fields as cost-accounting procedures, sales promotion, and production management. In the labor relations field its major function is compiling and distributing information and making suggestions on labor relations problems. The LNA has about two hundred member firms, drawn from the largest in the industry, which account for approximately one third of total industry employment. But this no longer measures the influence of the LNA in the field of industrial relations.

National Association of Photo-Lithographers. Apparently the genesis of the National Association of Photo-Lithographers was in a group of commercial lithographers in New York City who gathered together as early as 1927.[8] The executive vice-president until recently was Walter Soderstrom, and, unlike recent executive directors in the LNA, has been the dominant figure in deciding and implementing association activities. In several interviews the National Association of Photo-Lithographers was referred to as "Walter Soderstrom's outfit." The association takes even less active a part in labor relations activities than the LNA, although Soderstrom was active in the New York negotiations with Local 1 of the lithographers' union before choosing to spend full time on his association job. The key feature of the National Association of Photo-Lithographers is personalized service to members, plus the normal functions of providing uniform cost systems with a unique and quite useful aid for building hourly cost figures for separate cost centers in the plant. Like the LNA, the photo-lithographers hold annual conventions which maintain the atmosphere of a "working" convention more effectively than the LNA's. No NAPL speaker has ever promised to time his speech to allow for eighteen holes of golf before lunch.

The major venture of the National Association of Photo-Lithographers in labor relations was in the war and postwar period when it cooperated with the LNA and the lithographers' union in setting up mechanisms for labor-management cooperation, notably the Joint Lithographic Advisory Council. This experiment was abandoned, and since that time the association has not ventured into direct contact with the union.

Printing Industry of America, Inc. PIA is the successor to United Typothetae. The new organization has emphasized that PIA is different from the old United Typothetae:

> Printing Industry of America was *created* by the local associations; the locals were not created by the national as subordinate subdivisions thereof . . . As long as Printing Industry of America continues to be an organization by, as well as for, the local associations and their members, it will remain close to and responsive to the needs of the industry, and will accomplish the objectives for which it was organized.[9]

PIA has a larger staff than the other associations because of its larger membership, the majority of which are predominantly letterpress plants.

From its inception in 1945 PIA took pains to refer to "printing and lithographing" as its field of interest and in 1957 added to its staff George Mattson, a well-known lithographic industrial relations consultant. Two factors make Printing Industry of America the most important of the national associations as far as collective bargaining in lithography is concerned. The use of the lithographic process, as already noted, is expanding faster in the combination plants than in the straight lithographic plants, and combination plants make up the bulk of PIA lithographic membership. Second, PIA is fundamentally a group of local trade associations, and in the form of collective bargaining employed currently, these are the only trade associations that count for much. The Lithographers National Association (LNA) has said that "It works hand in hand with such of these [local organizations] as also recognized, reciprocally, the imperative need of a national organization fully equipped to deal with problems nationwide in scope." [10] But when local lithographer groups do recognize the need for a national affiliation, they often turn to PIA rather than LNA. A number of the major local associations remain independent of any affiliation. NAPL's Walter Soderstrom maintains friendly relations with local trade association executives, but the function of the NAPL is too closely tied to direct servicing of members to be carried out successfully through local associations.

Unlike the other national associations, Printing Industry of America does have contact with an international union and carries on direct negotiations. These relations are limited to a single union, the International Printing Pressmen and Assistants Union (IPP&AU), and relate to two subjects — a national arbitration agreement and a separate national apprentice agreement.[11] The negotiations are conducted by the union employers' section, and have no relation to the other wing of PIA, the master printers' (nonunion) section. Not all local agreements include the acceptance of these agreements, but it remains significant that national agreements are still possible on some issues in the printing industry. In addition to this direct connection with collective bargaining, the union employers' section carries on the normal data-gathering and -distributing function in the field of industrial relations.

Each of the three associations specializes in different types of services. The Lithographers National Association offers aids which are geared to the top management of a firm of 100 to 500 people; the National Association of Photo-Lithographers specializes in providing on-the-spot assistance and tailor-made accounting procedures; the Printing Industry of America offers services which capitalize on the network of local associations in reaching member firms.

There is, therefore, a difference in emphasis in each of the associations and in general a different "typical" member firm. But to a great extent the services do overlap, and there have been many efforts to bring all national lithographic trade associations under one roof. LNA has been successful in attracting several of the product-market associations to it, but not the photo-lithographers. As more firms become combination plants, there is an increasing rationality in making Printing Industry of America the single association. Recent efforts to join LNA to PIA as a semi-autonomous unit have failed, and there is no great likelihood that "Walter Soderstrom's outfit" will wish to subordinate itself to anybody. Like unions and their members, the associations and member firms do not always have identical interests.

There have been limited instances of cooperation at the association level, notably between the Lithographers National Association and the National Association of Photo-Lithographers in the early postwar period, and between the LNA and Printing Industry of America more recently. But as the type of firm from which each group draws its membership becomes more closely identical, the possibilities of friction between association executives increase.[12] The situation now parallels the relations between the five craft unions in the lithographic industry before 1915. There is agreement that amalgamation would be a good thing in principle and areas in which guarded cooperation is possible. But new faces and new problems will have to arise before "one big association" in the graphic arts can become a reality.

Canadian Lithographers Association. The situation in Canada is quite different from that in the United States, but closely resembles the state of management organization in this country a generation ago. Collective bargaining between the Amalgamated Lithographers of America and the present association appears to have begun in the mid-1930's,[13] and bargaining has been at a national level since that time in spite of the fact that the legal framework for bargaining in Canada tends to favor bargaining arrangements made at a provincial level.[14] The employers are a tightly knit group, and, like their American counterpart — the LNA — include substantially all the large lithographic firms, though a lesser percentage of small firms. The dispersion of the process has not proceeded as rapidly in Canada as it has in the United States, and this is one of the reasons the association has been able to provide a single voice of management in its relations with the union. The more active distaste for unionism which has been charged against Canadian employers generally[15] seems to hold true for employing lithographers and may provide additional common ground for action against the union.

It is possible that the Canadian Lithographers Association will go the way of the LNA. In the first contract signed in 1935, about 80 per cent of the membership of Amalgamated Lithographers of America in the five eastern Canadian locals was covered by the contract.[16] By 1951 this coverage was down to 65 per cent and by 1957 to 50 per cent.[17] There is, however, no evidence that industry-wide bargaining has become less stable as a result of this trend.

Local associations. All collective bargaining arrangements in lithography are made at the local level. Bargaining is occasionally on a single-firm basis, but the more common arrangement is for several lithographic employers to choose a single negotiating committee and conduct one negotiation with the union.

The group conducting the negotiations may be an informal group of employers, as exists in Buffalo.[18] In this city there are perhaps a half-dozen shops that may be classed as color or publications lithographers — two of which are still members of the LNA. The group has no name, no officers, and no dues — negotiating expenses being shared between the participating firms. There is, however, a local association of the Printing Industry of America in Buffalo, whose executive vice-president represents the combination shops in the negotiations as well as several straight lithographers, some of whom were formerly LNA members. This is a recent development, and the impression gained from the president of the lithographers' union is that the PIA-represented wing of the employer group is growing in importance. The recent success of the lithographers' union in taking over bargaining rights for lithographers in six combination shops from the printing pressmen's union has not only strengthened the lithographers' union, but the PIA wing of the employers' group.

Another type of bargaining arrangement is found in New York City, where the Metropolitan Lithographers Association represents the major employers in negotiations.[19] The association carries on certain sideline activities, such as arranging the joint sale of waste paper, but these are minor. Its reason for existence is to negotiate with Local 1 of the amalgamated. Although the Metropolitan Lithographers Association has less than a third of all lithographic plants in the New York City area among its members, these include all the large ones, and there is no other important contract for lithographic workmen in the New York City area that is not based on the association contract. The number of employer members is in any event a poor measure of an association's strength. In San Francisco, for example, only 15 of some 150-odd firms which negotiate with the amalgamated are members of the employers' association, but these 15 firms have over two thirds the total press capacity. This is the typical pattern in major cities.

The Printers' League Section of the New York Employing Printers' Association bargains with the International Printing Pressmen and Assistant's Union (IPP&AU) for offset workers in that union, and although rates are lower and hours longer (being tied to letterpress rates), it was estimated by the key negotiator for the printers' league that in the late 1950's less than 600 employees on offset presses and preparatory operations were represented by the IPP&AU in New York, substantially all of these among the low-skill classifications.

The Metropolitan Lithographers Association employs an attorney to assist it in negotiations, in day-to-day contract implementation, and to represent the association on the several programs administered jointly by Local 1 of the amalgamated and the employers. Bargaining, however, is still done by men who are themselves employers; the attorney is a key man in the organization and more important than the "executive director," but he is quite unjustly accused by many employers outside New York of being responsible for wage rates and working conditions which, taken as a whole, exceed anything else in the industry.

In some cities one firm may dominate all others and may be the only lithographic employer of importance in the city. Employer representatives in these cities are uniformly professionals, and are free from the multiple pressures familiar to the employer spokesman in group negotiations. There is only one set of markets, one set of production problems, and a single authority on the employer side, and the relative costs of a strike and a settlement on union terms will be weighed on a single set of scales. For these reasons bargaining is quite different in such cities from what it is in the larger centers, and the following "typical" bargaining arrangements are least applicable to them.

The Structure of Employer Bargaining Units

It would be difficult and unproductive to describe in detail the different types of arrangements that determine who sits on the employer side of the bargaining table, and where the crucial decisions are made for the employers. The Boston employers' group may be taken as representative of the industry, and the following description will draw primarily on that group.

The employer representative in Boston is the Graphic Arts Institute of New England, one of the many PIA-affiliated local associations that conduct negotiations with all printing unions in a city, including the Amalgamated Lithographers of America.[20] The fifteen signers of the master contract employ about three quarters of ALA Local 3's membership. The firms make up the Lithographic Division of the Contract Employers Group, and all negotiations are conducted by employer representatives, with the local trade association executive occupying a unique position as a member of the employer bargaining group, whose detailed notes of the bargaining sessions are also distributed to and relied upon by the union committee. Negotiations have no particular connection with the PIA; Forbes Litho is a member of the Lithographers National Association; Buck Printing and A. T. Howard are members of the National Association of Photo-Lithographers, so that the facilities and information provided by all three national associations are available to the bargaining group.

The Boston employers, like the Boston ALA local, had long been split between Forbes Lithograph on the one hand and the rest of the shops on the other. Forbes is a large color house which was one of the founding members of the National Lithographers' Association in 1888, having the largest employment of any of the fifty member firms at the time. From 1922 to 1937 Forbes was a nonunion shop, but from that time to the present has had some sort of collective bargaining arrangement with the union. Not until 1954, after a stiff settlement in 1953 and the application by Local 3 of successful whipsawing tactics between Forbes and the Contract Employers Group, did Forbes agree to joint negotiations. Prior to that, Forbes management negotiated independently of the other Boston employers because of the supposed difference in their production problems and the markets in which they competed.

The Contract Employers Group had been formed in 1949 with five firms, to avoid whipsawing by the union. As the number of shops under contract increased, it became apparent that a precedent set in one shop, where the point at issue was unimportant, could be used with considerable effect in another where it was by no means unimportant. The addition of Forbes in 1954 left no important shop outside the group, and as the number of shops under contract to the ALA has increased, so has the size of the Contract Employers Group.

Of the fifteen firms that made up the employers' group in 1959, the owner-managers of eight of them were by general agreement considered to be unsuited for negotiations. It is probable that they themselves would have disputed this only if there were some other good reason that would have permanently excused them from having to act on the negotiating committee. Of the seven remaining, two were non family-owned and -operated firms, each having an executive whose major concern was industrial relations. Forbes was one; the other was the Rust Craft Greeting Card Division of United Printers and Publishers. Because of their size and the ability of their staff men, these two firms had always been represented on the negotiating committee. In the 1958 negotiations two other firms were represented by their owner-managers on the negotiating committee, both of them combination shops which are now predominantly litho. The two trade shops in the group were not represented, nor were the third- and fourth-largest organized litho firms in the area, though all were in the Contract Employers Group.

The typical method of choosing employer bargaining representatives is for employers active as representatives in the past to decide who among the managers of firms to be represented can safely be allowed to sit at the bargaining table and contribute something, and to add these persons to those who are willing to continue or who insist on continuing as a part of the negotiating team. The names are then formally presented and accepted as the bargaining representatives of all firms in the Contract Employers Group. In the past, arrangements were made to give any employer not on the committee permission to attend bargaining sessions. This was necessary, in the words of one employer, to allay the mistrust that existed among the employers.[21] Inasmuch as it interfered with negotiations to have a strange employer dropping in from time to time, the practice was discontinued.

The assembling of data and development of the employers' bargaining position is done primarily by the two or three active members of the committee. The executive director of the Graphic Arts Institute makes available to the committee the material which PIA supplies and any other available to him, but most of the work associated with relating such data to the specific bargaining problem at hand falls on those members of the committee willing to do the work. Not a great deal of weight is placed on rates which are paid in other cities, both because Forbes is the only firm which sells extensively outside the Boston area, and because Boston rates are significantly under those of New York, the most important competitor. A good deal more stress is placed on the hourly labor costs of unorganized local firms, but at least in the recent negotiations no effort was made to support the arguments in this area by gathering factual data of orders lost, the extent of underquoting, and so forth, which was done in recent New York negotiations. The single case in which a statistical presentation was used effectively was in support of an argument relevant to only one company, and to some extent opposed by another member of the negotiating team.

The method of reporting back to the employer group is informal, commonly talking with individual members over the telephone about high spots of the negotiations. Minutes of the negotiation meetings are also sent to all employer members. Decisions as to the amount and form in which concessions will be given rest entirely with the committee and are not formally or otherwise presented to the employers' group. The final settlement is typically on the basis of an employer's last offer "which exceeds the authority we have been given by our group." Employers have been known to refuse to go along with such a settlement, but the alternative is a strike with all the other firms working, which cannot be classed as a genuine alternative. Limits are placed on the discretion of the committee by the fact that they are themselves major employers. But the firm which has an operation peculiar to that firm alone is well advised to participate actively in committee deliberations if the operation is a subject of any union proposals.

It is to be expected that the effectiveness of the employer bargaining team will not be outstanding. The bargainers are for the most part company managers who rose to their present position through skills which do not transfer readily to the bargaining table. They have full-time jobs to which the annual or biennial negotiations are an added burden, which becomes considerable if the negotiations drag out and the issues demand much consultation or cause friction between employers or tension between the employers and the union. The phrase "hard negotiations" seldom describes anything exciting; it means meeting after meeting in which the same unresolved issues come up to be argued intensely in much the same terms, and possibly a few new examples. It requires listening to a man whom one may dislike and for whom one may have little respect, but sitting there just the same, waiting to refute the statement he has repeated for the third time. People accustomed to giving directions are not well suited to this sort of thing. Notes of an interview with an employer suggest the sort of problems that in fact did arise in negotiations in his city:

[The ALA local president] is a stupid and difficult person to get along with. The ALA is strong here, but he is useless, has absolutely no understanding of economics or the employers problems . . . Their demands last time were utterly ridiculous. We were sitting there for five weeks with them asking for 25 cents when they knew they wouldn't get it. They knew it, and we knew it, and still we had to sit there and listen to the asinine arguments. They think they're big shots — wanted guaranteed holidays. Why the hell should they be paid for that? They wanted a fixed starting rate for general workers. Why the hell should I have to have a fixed rate for that? I have to pay what the market calls for. That's none of their business. They delay negotiations just to get supper money. I think they like to negotiate, they love to waste the time and act important.[22]

Not surprisingly, this comment was made by the trade association executive of the same city:

There is one outstanding difference between union and employers' organizations which accounts for the strength of the former and the weakness of the latter at the bargaining table. Whatever their differences may be or how bitterly they are contested, once the policy has been adopted, either nationally or locally, it is the union policy. Bargaining strength is the result of their union on other matters. On the other hand, only the threat of complete union domination drives employers into a pseudo organization for bargaining only. Those representing the employers' group know, even while bargaining, that their principals are only giving lip service to the principle of unity and will follow what is agreed to only as long as they must or it is to their obvious interest to do so.

On occasion the ineptness of owner-managers has achieved a certain notoriety. The following instance, which occurred during some hard negotiations in another city, is such a case:

[He] reviewed his background from the time his parents came to this country as immigrants . . . He closed by saying that the employers have not got the money and took a turnip out of a paper bag and explained that no blood can be squeezed out of it. He cautioned Local —— that if they drive the employers too far they may drive them away from offset and into roto-gravure.[23]

Employers such as these have led to the increasing use of professionals at the bargaining table. These professionals need not be hired by the association; they can be the industrial relations officers of a firm, but there is a growing consensus that line managers ought to be kept away from the bargaining table. Many of them realize this. As one veteran of several negotiations pointed out, "A staff man finds it easier to say no. An employer can say he has to go back and consult, but if he is the top man in the company who will believe him?"[24] For employers too small to hire their own staff man, there is a natural preference for a person unconnected with any other firm to represent them at the bargaining table, and in the last few years there has been a sharp increase in the number of lithographic negotiations in which a leading member and occasionally the spokesman on the employers' committee is a full-time trade association executive.

The advantage of such an arrangement is not solely in the replacement of owner-managers in bargaining. With the increase in the number of locals having paid, full-time officers, there is a growing need to have some-one on the management side who can devote full time to the problem of preparing for negotiations, administering the contract, or at least coordinat-ing its administration, so that the actions of one employer do not create an embarrassing precedent for others.

The Amalgamated Lithographers of America

Lithographers are found in many unions, though one, the Amalgamated Lithographers of America, is clearly preeminent. The United Steelworkers have a number of metal lithographers organized in can plants, the United Automobile Workers have some in auto and parts plants, United Paper-makers and Paperworkers have others in boxboard plants. Although these unions affect the lithographic industrial relations system, they are not actors in it. Rather, their strength and the nature of the industry they organize determine that the lithographers who are their members will have working rules determined in a context quite different from that of the lithographic industrial relations system.

Within the printing industry, there are three unions other than the ALA which organize lithographers — the International Printing Pressmen and Assistants Union (IPP&AU), the International Photo-Engravers Union (IPEU) and the International Typographical Union (ITU). The impor-tance of these three unions is closely related to the lithographic skill groups they organize, and a discussion of their role will follow the description of the workforce in Chapter V.

Membership of the ALA.[25] The Amalgamated Lithographers of America is the only union whose major interest is in the lithographic industry.

In 1959 its 36,000 members were about 50 per cent skilled journeymen, 20 per cent less-skilled journeymen (such as feeders), 15 per cent apprentices, and 15 per cent "general workers," primarily paper handlers and finishing department help. During the late 1930's this latter group of members ac-counted for nearly one third of the membership, but the more recent policy of the union has been to discourage locals from taking in more of the un-skilled workers, and the proportion of membership made up by general workers has declined steadily. Even in unit contests with other craft unions, the ALA typically has sought a craft unit narrower than the opposing union desires. Wide variations do exist in the importance of general workers within locals. For example, in 1959, Rochester had nearly as many general workers as journeymen; New York, one third as many; and Chicago one twentieth as many.

Estimates of degree of organization in the ALA among the skilled workers require the joint use of ALA and Census data which introduces some error into the figures.[26] Using this material, the degree of organization in six major metropolitan areas in 1957 is given below:

New York	86
Chicago	92
San Francisco	85
Philadelphia	65
St. Louis	76
Los Angeles	56
Cincinnati	81

One cannot expand these figures into an estimate for the total degree of organization by the ALA, since, in general, organization is less complete in smaller cities. For the late 1950's perhaps 60 per cent would be a reasonable estimate for the ALA's over-all degree of organization in the skilled category of workers.

The ninety-three locals of the ALA are spread among forty-two states and six Canadian provinces. Local 1, New York, has gradually lost its commanding position in the union, dropping from 36 per cent of total membership in 1951 to 21 per cent currently, though its membership still exceeds the combined membership of the seventy-five smallest locals. New York's relative decline has lessened the importance of the large cities taken together, but if that city is excluded from the group, the other large locals have grown at about the same rate as the union as a whole. Average local size has increased over the period 1915–1960 from 123 to 387, yet half the locals today have less than 100 members. Size distribution in 1959 is shown in Table 3.

While the majority of the small locals are in towns having no large lithographic firms, a number are so-called one-shop locals. Bennington, Vermont; Niles, Michigan; and Ashland, Ohio, have ALA locals only because there is one large lithographic shop serving a national market in each of them. Other locals of this sort could be listed, and one must conclude that a majority of the lithographic workmen serving local markets

Table 3. Distribution in size of locals, 1959.

Local size	Number of locals	Per cent of membership
0–99 members	47	5
100–499 members	9	17
500–1499 members	13	32
1500 and over	4	45
	93	10

outside the major cities are either unorganized, or organized by unions other than the ALA.

Historically, the growth of the union has not been even, but does reflect the long-term growth of the union movement in America generally. The following list gives ALA membership figures at five-year intervals (for December 31 each year) since amalgamation:

1915	4,199	1940	12,849
1920	7,619	1945	16,429
1925	5,466	1950	25,232
1930	5,717	1955	30,453
1935	6,877	1960	37,959

Growth of the ALA since 1950 has been steady, but not spectacular. The 39 per cent increase (to 1958) may be compared with a 20 per cent increase of total United States union membership on the one hand, but with a 57 per cent growth in industry employment on the other.[27] The growth of the union is not keeping pace with the growth of its claimed jurisdiction. Its coverage of the industry is less complete than it was in the 1930's, and is probably less than at any time since the early 1900's, excluding periods following major strikes.

The Structure of the Union

There are four organizational levels in the union: the international, the region, the local, and the shop. The international and local organizations are the ones of outstanding importance.

The international. The formal organization of the international is shown on the chart. The international council dominates the chart as, in practice, it does the union. The president, vice-presidents, and secretary-treasurer are each members of the policymaking international council, and are also executive officers. They are the only officers who have a union-wide electorate. Ten of the seventeen members of the council have the title of international councillor — an unpaid office. They are elected by a majority vote of the regions they represent, and are usually the presidents of leading locals.[28]

In times past, the international president has been the true leader of the union, but a succession of weak presidents and a challenge whose solution rested in the application of legal expertise to organizing, negotiating, and jurisdictional strategy put the union's general counsel in the key position. This still left a void, inasmuch as the general counsel could not represent a political center of gravity. This fact, combined with the existence of such normal tensions in the union as large local versus small local, Chicago versus New York, electing the majority of council members on a regional basis and the rest on a national basis, has led to a situation where political power has been centered in the regionally elected international councillors.

The most important of the international councillors are the presidents of the large locals, notably the New York president. Because of greater face-to-face contact with members of their locals, and less opportunity for organized opposition to develop, these officers can lead their locals in a way that would be impossible for the international president to lead the international. Unlike the international officers, their regional constituency insulates them from political opposition arising in other parts of the country. For example, a central region councillor will normally pick up a few votes if he is known to be opposed to the New York local leadership for one reason or another. Other than this, the international councillors depend on their own vote-getting powers, and at least one or two contested spots are common. There is no good reason, for example, why the president of San Francisco rather than Los Angeles, or the president of Minneapolis-St. Paul rather than the president of St. Louis should be on the council, and a poorly handled negotiation can sometimes be the factor which determines an election.

The international president is constitutionally almost powerless. Unless he is able to count on the support of the council, he is able to initiate no action on his own, and may indeed be forced to take actions repugnant to him. To cite one recent example, ex-President Canary while still in office wished to support the drive of Chicago, his home local, to move the international office to the midwest. This is an ancient question in the ALA, having been raised in nearly every convention since the first in 1917. But

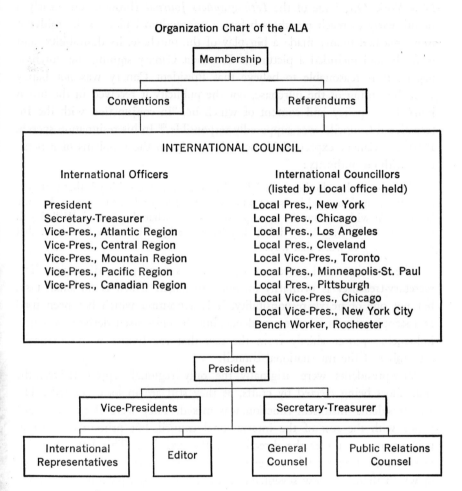

Organization Chart of the ALA

Membership

Conventions

Referendums

INTERNATIONAL COUNCIL

International Officers

President
Secretary-Treasurer
Vice-Pres., Atlantic Region
Vice-Pres., Central Region
Vice-Pres., Mountain Region
Vice-Pres., Pacific Region
Vice-Pres., Canadian Region

International Councillors
(listed by Local office held)

Local Pres., New York
Local Pres., Chicago
Local Pres., Los Angeles
Local Pres., Cleveland
Local Vice-Pres., Toronto
Local Pres., Minneapolis-St. Paul
Local Pres., Pittsburgh
Local Vice-Pres., Chicago
Local Vice-Pres., New York City
Bench Worker, Rochester

President

Vice-Presidents

Secretary-Treasurer

International
Representatives

Editor

General
Counsel

Public Relations
Counsel

Source: ALA, *Constitution,* 1958 revision, International Councillors as of December 31, 1958.

Fig. 1

while the political maneuvering was going on, the council, a majority of which opposed the move to the midwest, exercised the authority given to it by an earlier convention and bought an office building in downtown New York. One issue of the *Lithographers Journal* (brought out nearly a month early to reach members before a referendum ballot on the midwest move reached them) made a big play of the purchase, its desirability, and so forth, and included a picture of President Canary signing the purchase papers. It is reasonable to believe that President Canary was not happy with the timing of the purchase, nor the publicity it received in the union journal, the policy and content of which he, "in conjunction with the International Council," was supposedly responsible.[29] In his resignation speech, President Canary expressed himself vigorously on the problems of a president without authority:

Now then, what do they want? Do they want a guy on this job that just goes along with anything and he is not allowed to have a mind of his own or, because he is not in with a group he is no good — is that what they want? Well, I will never work that way. Why didn't they let me resign when they found out that I wasn't going to work that way?[30]

The functions and powers of other officers require less discussion. The secretary-treasurer is primarily an office manager, which is no small task, but not one of heavy responsibility. It is a position which has been used as a steppingstone to a vice-presidency, but the office itself derives its policy-influencing power chiefly from the fact that the holder is automatically a member of the international council.

Vice-presidents were originally the only regional representatives, the councillors being elected by crafts, as they still are in local councils. The vice-presidents' historical function was mixed. They were elected to make policy with the rest of the international council, but primarily to act as paid organizers under the direction of the president. This latter function was to a great extent taken over by appointed international representatives, of which there are now seventeen on the international payroll. As a result, the vice-presidents became either elder statesmen or prestige organizers, in either event an expensive luxury for a union of 36,000 members. Only in Canada, where bargaining is on a regional basis, did the vice-presidents have an important role to play. Organizing is a young man's job, and vice-presidents, once elected, were usually safe for life. Presently the ALA is shifting from a regional to a functional assignment of responsibilities (one vice-president in charge of negotiations, another in charge of organizing, and

55 SCREEN

85 SCREEN

120 SCREEN

150 SCREEN

Figure 2: Illustration of picture with different dot sizes.

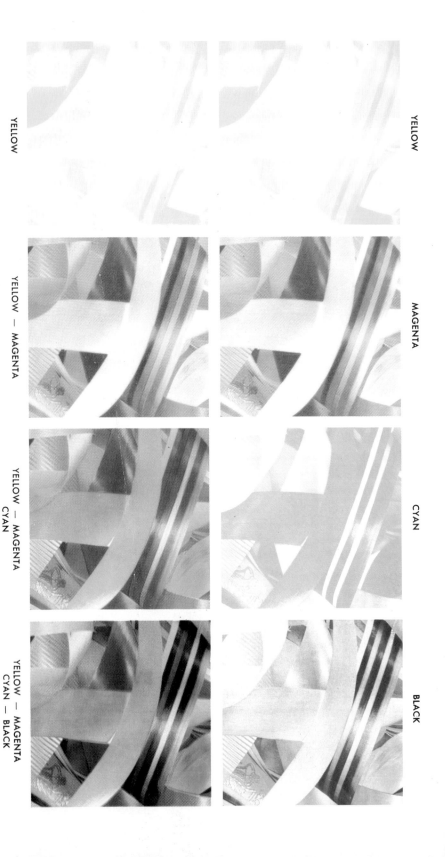

YELLOW

MAGENTA

CYAN

BLACK

YELLOW

YELLOW — MAGENTA

YELLOW — MAGENTA
CYAN

YELLOW — MAGENTA
CYAN — BLACK

Figure 3. An example of four color, 150 line screen photo-lithography. Top row: proofs of each color plate. Bottom row: progressive proofs of each color plate. (Reproduced by courtesy of Kimberly-Clark Corporation whose color separated photo-lithographic negatives were used by the Nimrod Press for platemaking and printing.)

so forth), but it is too early to say whether this represents a substantial change in administrative practice. The editor of the *Lithographers Journal* was once an officer of the union and a voting member of the council. The present editor is a paid employee, as was the former public relations counsel, and the dozen or so office workers.

In most cases the lines of authority from the elected to the hired official are not clear. The editor is formally under the direction of the president, but must be responsive to the international council. The office staff is not faced with the same split between formal and effective authority, primarily because in the past several years the secretary-treasurer was the only international officer to make the international office his headquarters. The problem is more acute in the case of the international representatives, full-time organizers who are appointed, not elected, but who are chosen from the ranks of successful local union officers. Until recently the representatives were under the international vice-president of the region in which they served, under the general direction of the international president, and responsible to the local president in whose jurisdiction they were working. In practice, their activities were also watched closely by the international council. Perhaps more than any other active group in the union, the international representatives suffer from multiple and conflicting lines of authority.

General counsel. The ALA has been described by one of its critics as the only union in America that does not even go through the motions of electing the man who runs it.[31] The reference is to Benjamin Robinson, general counsel for the ALA since the mid-1930's. There is no question that Robinson in times past was the power behind the throne and is still influential in the councils of the union. The explanation of his power lies primarily in the jurisdictional controversy.

The security of the union is tied closely to the preservation of its jurisdictional position before the NLRB, and the general counsel has worked skillfully to preserve that position. The international council firmly believes in the importance of preserving this position, and as a result its decisions rest heavily on the advice of a man whose ability in this field is not questioned by his most vigorous opponents. He has faced opposition within the union, but his influence continues with those elected policy makers who believe in the wisdom of his counsel, and who respect the experience of the man who gives it.

It may be helpful to review the ALA international structure briefly. The formal organization of the ALA is not materially different from many other unions. The quality which makes this union distinctive is the informal distribution of power within the formal organization.

There is a plural executive, the international council, which determines policy and initiates action. The president is an administrative officer who, like the secretary-treasurer and the five vice-presidents, reports to the council for orders. One of the key policy planners is a staff man, the general counsel, but he also is administratively responsible for a key activity — jurisdiction. Therefore his authority stems not only from his ability as a strategist but from his superior understanding of the legal tangle that surrounds the jurisdictional question.

Like all generalizations, these require some qualification. In recent years Benjamin Robinson's participation has not been so vigorous as it was in the late 1940's and early 1950's, and a major shift in international leadership in early 1960 may well cause further changes in the distribution of power in favor of the new president, Kenneth Brown. Since some international councillors have been among the most vocal critics of the way the union has been run, such a shift is likely.*

The local union. The local union is an important center of power in the union, both in matters concerning union politics and in employer relations. Although the international can refuse to ratify an agreement satisfactory to the local and thus deny the local strike benefits and the employer the use of the union label, such power has to be used sparingly. For in the final analysis, local bargaining means that the attitude of the membership in that local will be the key determinant of an "acceptable" bargain.

Like the international, each local has a council made up of the officers of the local and representatives of each craft, or, more accurately, each department.[32] There are several reasons, however, why the local council plays a much less active role in the affairs of the local than the international council does in the affairs of the international. Because the membership is not scattered as it is in the international, important decisions may be taken by majority vote at local meetings. The policy-making function is therefore not so important a function of the local council. Moreover, members of the local council earn their living at the trade; they are not, as are the majority of international council members, paid officials of the international or of local unions. Most important, the president of a local is a key figure in local

---

* In August 1962 Robinson was asked to cease representing the New York local. He refused to do so, and was then removed as general counsel.

politics, with a direct face-to face contact with the membership. Because the membership is able to exert pressure on the local president directly, there is less need for the representative democracy provided by the international council.

This in no way implies active participation by the membership in local affairs. Typically, the most important internal local problem among ALA locals interviewed was lack of member interest.[33] This requires most local presidents to take the initiative in nearly all union activities, and also permits them to do things the way they want to. One local president said that after several years of trying to share in the decision making, he got sick of hearing, "Well, what do you think?" and ran the show himself. This pattern of behavior was clearly acceptable to another local president:

If anyone complains about how slow things are going I say "How about you getting on the committee?" One guy raised quite an argument at a meeting, I told the shop chairman, "I want you to put that guy on the bargaining committee next time." That cooled him right down.[34]

The point of significance is not that presidents do things "their way," but that their way is acceptable to the membership. Membership apathy is a leadership rather than a membership problem. Other local presidents have more success with sharing responsibilities; and particularly in large locals such as New York, Chicago, and Toronto, there are competent second-level officers who are encouraged to take on responsibilities by the respective local presidents.

Although New York has had a paid president since before the turn of the century, only three or four locals had paid leadership up to the end of the 1930's. Boston, for example, established a full-time position in 1942, Toronto in 1945, Buffalo and Rochester after this time. Benjamin Robinson emphasized during interviews that many of what now are considered standard practices in local situations developed during the period of great expansion in the number of written contracts from 1937 through 1941, when the great majority of locals were led by men who had to work full time at the bench.

Following the war, the practice of paid leadership spread to most locals with three hundred to four hundred members or more; and in the mid-1950's the importance of full-time local leadership led the international to initiate a selective subsidy program to permit full-time officers for locals which needed them but could not afford them. In the 1958 Policy Conference this program was expanded to help locals, even with less than a hundred members, to put on a full-time officer whose major duty would be organizing. Currently thirty-two of the ninety-three locals have full-time officers.

The smallest unit of union organization is at the shop level. A shop delegate is a key figure in local affairs, both because he is usually the spokesman and to some extent the opinion leader in the shop, and because he can make things exceedingly difficult for the shop management if frictions arise between it and the workers. Although the "mutual government" which the employers proposed a half-century ago has been quite thoroughly implemented today there are still parts of the union constitution which are considered an unjust restriction when applied in "the passageway which is the trade." Among such clauses are the provisions controlling apprentice ratios, complement of man on presses, and overtime restrictions. In most cases the union constitution is viewed as a flexible document, containing both absolute requirements and optimistic goals. But nothing in the document explains which are which, and it is therefore possible for a shop steward bent on troublemaking to clothe his tactics with constitutional, if not contractual, sanction. Needless to say, the shop steward may also simplify managerial problems, and in practice this seems to be the more common case.

Arrangements for Bargaining

We have already noted the general rule of one negotiation for one city, with a group of employers being represented on one side of the table, and the local union on the other. The method of choosing the local union bargaining committee varies from city to city. In some they are elected from the floor; in others the president takes a more active role in choosing them, even to the extent of appointing the committee. In others the bargaining committee consists of the top officers and the shop delegates. The latter is usually true only when a single shop or at most two or three are represented in the negotiations. However chosen, the general rule is to approach the following ideal, in the order given, as nearly as possible: have a committee representing (1) each major branch of the trade, (2) each kind of shop (large, small, trade shop, combination shop, and so forth), (3) who are not blowhards, (4) whose opinions carry weight back in the shop, and (5) who are good thinkers.

Most union presidents want a committee large enough to be representative more than they want a committee small enough to be flexible. In some locals a compromise is reached by having a four- or five-man bargaining committee which is a part of a larger policy committee that does not attend negotiations, but meets once a week to review progress and thus shares the responsibility for the package when it is presented to the membership. In Boston, for example, the policy committee is made up of the local council including the officers of the local, and all shop delegates. In other locals the council alone may serve the same purpose.

Talking at the bargaining table is restricted primarily to the president. In all the locals interviewed the president was the head of the union committee and also its spokesman. In some locals where there is no paid president an international representative will take an active and continuing part in the negotiations, but even in these the representative is there at the request of the local president, who, if not better able to conduct the negotiations, is often better able to determine what issues are most important to the membership.

Often local priorities will not be the same as international priorities, a fact which puts the local president in a difficult position. He is usually under pressure from the membership to get a package that is long on money and short on everything else, and to get it without a strike. He is under pressure from employers and is usually better able to see the weight of their economic arguments than the membership, for he is continually reminded by them of the low-wage competition from shops he has failed to organize. And he is under pressure from the international to stand up like a man and "sell" the membership on what is good for them, and then go in and convince the employers that nothing else will be accepted. The ability of local presidents to do this varies widely, and depends not only on their powers of persuasion, but on the time they can spare for union activity, and the extent of their commitment to the union. A subsidization program which permits locals with as few as fifty members to have full-time officers is one of the means the international has adopted to improve the quality of local representation in bargaining. It is not yet possible to say whether it will be successful, but it does provide the union with something which some employer groups do not have, a person who can devote full time to the problems of improving and developing the bargaining position of the people he represents.

Chapter **III** ✑

The March of Technology

One of the most significant features of the lithographic industrial relations system is the dynamism and accompanying instability, initiated by the rapid pace of technological change. Over the years innovations have destroyed the industry's relative isolation by making the process an effective substitute for more traditional printing methods. Lithography is no longer limited to making the lithographs and gaudy posters of a bygone era; it is now a quick and economical method for producing books, handbills, maps, display material, labels, Christmas cards, brochures, and other printed material.

Historically, the lithographic process relied on the quality of a certain type of porous limestone to accept either water or a greasy ink on its prepared surface, but not both in the same area. The stone was then placed in a flat-bed press, dampened and inked, and then impressions were drawn from it. At the turn of the century, the cumbersome lithographic stone was gradually replaced by a thin metal sheet and the old flat-bed press by an offset press, so called because the thin metal plate, strapped to the plate cylinder, transferred the image to a rubber-blanketed cylinder which "offset" the image to the paper as the paper passed between the rubber blanket and the impression cylinder.

The offset press is the nub of a fifty-year-old jurisdictional controversy between the lithographers' union and the printing pressmen. The printing pressmen claim that "lithography" is by literal definition limited to printing from stone and that the offset press is merely one of the many types of printing presses which belong within their jurisdiction. The printing pressmen have remained literal purists in this repect for a half-century:

Lithography, or printing from stone, was a process altogether different from the offset process of our time, but served as a jumping-off place for the modern offset method.

If there can be said to be an overt place, a line of demarcation, this [the introduction of the rubber-blanketed third cylinder] was the time that lithography passed from the scene. Of course, other developments . . . hurried the death of lithography.[1]

Industry usage has not followed the pressmen. Outside of Pressmen's Home, Tennessee, few argue that "lithography" refers to occasional artists working on stone, and not to the fastest-growing segment of the graphic arts.

The printing pressman believed that the offset press would cut seriously into the letterpress market; they did not believe it would require lithographic craft skills to operate it. They were twenty-five years too early with their first prediction, and wrong on the second. One must thoroughly understand an innovation to predict its effects. The linotype, for example, broadened the typographic market to the benefit of other printing trades as well as the typesetters but did require a basic change in craft skills. On occasion the offset press and linotype have been compared, yet one could not find two innovations with more contrasting effects. To mention only one, the linotype helped split the typographic union into five printing unions, while the offset press helped unite five lithographic unions into one.

It is altogether useless to say "innovations are important," and then proceed to generalities about the effects they *may* have. Innovations may increase skill requirements or decrease them; may cause jurisdictional disputes or eliminate them; may make a firm of five employees profitable or require five hundred. One needs to look specifically at the production process and the changes that are set in train by the introduction of a new method to assess an innovation's impact on the industrial relations system.

This is particularly true in lithography. For example, the introduction of the offset press led directly to the claim of jurisdiction by the printing pressmen and the photo-engravers in 1913, but the process remained sufficiently unique to make their victory in the councils of the AFL a failure in fact. When the struggle was rejoined before the NLRB thirty-five years later, the issue hung on such points as the difference between a deep-etch plate and a photo-engraving, and whether or not the lithographic platemaker required a knowledge of chemicals. A current example also illustrates the impact of technology. Will the lithographic industry lose its separate identity within the printing industry? One must look to the future of paper and presensitized plates, the simplified offset press, and photographic typesetting for at least part of the answer.

The Lithographic Process

All printing processes involve five steps: (1) obtaining or creating copy, (2) changing the copy so that it can be used for printing, (3) making a printing surface, the plate, (4) producing the product, and (5) cutting, folding, binding, or otherwise finishing the product.

The first and fifth steps are not a part of the central process for the purpose of this study. Members of the creative art department are now considered a part of the professional group, and binding and other finishing operations are separable because the skills are quite different from the three central lithographic operations. Moreover, they are common to all printing processes and are normally organized by a separate labor organization.

Copy preparation. The ease with which simple copy may be prepared is one of the unique features of the lithographic process. Line drawings or text matter, for example, need only to be photographed, and the resulting negative used directly to make the plate. Copy may be enlarged or reduced photographically with equal facility, and the low cost of repeating the same advertising message or theme in sizes appropriate to *Coronet, Life,* a counter display, or a billboard gives the lithographic process a substantial edge in supplying this market. If the text is still in manuscript form it can be "set" directly on film by one of the several photographic typesetting machines now on the market. When type forms are available the transfer to film can be either by printing a proof copy and photographing it, or by a recently introduced process which photographs the type face directly, producing sharply defined characters even from worn type. However, the ease with which copy prepared for letterpress may be adapted for lithographic use is not yet a two-way street, to the considerable disadvantage of letterpress.[2]

When reproducing photographs or other copy in which there is continuous gradation from light to dark (called "tone" copy), direct photographic transfer is not possible, since the amount of ink transferred to the paper in both the letterpress and lithographic process is the same in all parts of the image. Such copy must be photographed through a fine screen, the image appearing on the negative as a series of minute dots as many as 125,000 per square inch, which are large or small according to the darkness or lightness of the original image.

If a finished sheet is to include several units of copy, such copy must be precisely positioned prior to making the plate. This step is called stripping. Complex stripping operations require the use of additional photographic techniques, and also call for considerable skill and close coordination between the cameraman and stripper. The key requirement in stripping is to ensure that copy will print in the correct place on the paper; that is, that all copy will be in "register."

Printing in more than one color adds complexity to the stripping operation. When printing one color, units of copy must be assembled for exposure to a single plate; with several colors, however, different plates for each color, mounted on successive units of a multicolor press or each in turn on a one-color press, must overprint each other precisely.

Because camera lenses and filters are not perfect and because inks available are much less so, correction is needed on each color separation negative. Recently, the amount of manual correction necessary — the job of the "dot etcher" — has been substantially reduced by masking techniques in which additional photographic negatives or positives are used as an overlay to enlarge selectively or reduce dot size and, therefore, the amount of color deposited during the printing operation. Much more than dot size however needs to be considered in the preparatory stage. Other factors such as angling, size (mesh) of the screen, or one of the many variables in lens, filter, lighting, emulsion, and others may have to be adjusted to modify the separation negative and achieve the desired results.

Other methods of producing color work can be used which reduce the number of steps and the amount of human judgment required by standardizing procedures and materials, but at the sacrifice of some quality. The effect of these methods is not to reduce the cost of existing products, but to broaden the market which the lithographic process can serve.[3]

A recent and important innovation is the electronic scanner, a device operating on essentially the same principles as the television camera and makes partially corrected separation negatives from color slides. It has a high-output capacity, variously estimated at ten to fifty times that of a conventional process camera.[4] Making color separations by cameras is still more efficient for some types of work, and even on scanned separations some color correction is usually necessary. In spite of this, the scanner may put the cameraman and the dot etcher in the same category as the platemaker. The skills are still required, but in much smaller proportion than before.

Platemaking. Platemaking involves two steps — preparing the plate to receive the image and placing the image on the plate. During the 1930's there were important improvements in plate preparation which lengthened the life and improved the predictability of plates. Nearly all these improved methods were developed by the Lithographic Technical Foundation (LTF), an organization sponsored initially by the major employers' association and financed by lithographic firms to conduct research and educational programs for the industry. The LTF has served the industry well and is directly responsible for many of the most important cost-reducing innovations introduced in the last thirty years.

The postwar period has seen the elimination of plate preparation in many lithographic shops through the purchase of presensitized plates which need no preparation for exposure. Some can be purchased for less then ten cents a plate, good for about 750 impressions, while other and more expensive ones can be trusted for 100,000 impressions and more. For smaller firms, this one development has dispensed with the need for two bulky pieces of equipment, the plate-graining machine and the plate whirler, as well as the supply of various chemicals and sheets of metal. Paper-base presensitized plates have been developed which may be "made" directly on a typewriter, or by some other means that transfers the image by pressure rather than by photo-mechanical methods.

The method of transferring the image to the plate has also changed over the years. Until the late 1920's and early 1930's the standard method was for a hand transferrer, using a sheet of transfer paper, to pick up an image from an engraved stone and transfer it to the plate. No photo-mechanics and therefore no light-sensitive surfaces were used. The standard method today is to transfer the image in much the same way a contact print is made. The photographic transparency is pressed tightly against the light-sensitive surface of the plate in a vacuum frame and then exposed to light. If the plate is to carry the same image many times, as in the case of cigar labels, the vacuum frame is replaced by a photo-composing machine in which the negative or positive can be exposed in one part of the plate, moved to the next predetermined position, exposed again, and so on. Introduced in 1912, this machine was first referred to in union discussions as the "transfer machine," since it effectively displaced the hand transferrer. Photo-composing machines are now available in which the step-and-repeat operation is controlled by a punched tape. Exact registration, formerly the responsibility of the machine operator, is transferred by this machine to the planning department. It is the photo-composing machine which gives lithography its cost advantage over other methods of printing when multiple imposition of images on a single plate is required.

Press operation. Changes in the production process have also been important in the actual printing operation. It is necessary to remember that the feature which ties all lithographic printing together is the use of an impression surface — the plate — in which both the image and nonimage areas are in contact with the print-receiving surface. This surface may be paper, cardboard, metal, or cloth, the press may fill a good-sized room or be no larger than a duplicating machine, but the operating principles will be identical, and press construction surprisingly similar. The typical press will have three cylinders, a plate cylinder, around which the plate is fastened, an impression cylinder: and a blanket cylinder between the two, the rubber surface of which picks up the image from the plate cylinder and transfers it to the paper or other print-receiving surface as it passes between the blanket cylinder and the impression cylinder.

The plate is inked by an inking system similar to that of relief presses, but the lithographic press also has a dampening system because nonimage areas of the plate must be kept wet, or else these areas will pick up ink from the inking rollers.

An excerpt from a technical forum suggests some of the variables involved in the dampening system, quite apart from the range of problems flowing from the chemical properties of the solution itself:

> At the laboratory, we are still working on this problem of trying to find out just what happens in the dampening operation. We discovered here several years ago that one of the big troubles in dampening was that when you transfer water to the plate, we say that the water should not wet the image areas. Well, actually it does. Water that transfers to the nonprinting areas, being water receptive, just distributes all over the plate. But the water that comes in contact with the image has no place to go, so it breaks up into little droplets, about .001″ in diameter . . . So we're making a study of these water droplets and just what forms them, and maybe try to find some ways of reducing them.[5]

Similar problems arise with the ink, which requires a more complicated mechanical system than the dampening solution. This complexity is required by the multiple functions of the system, the most important of which is to permit the pressman to lay a uniformly thin film of ink on the image surface of the plate. Since 1958 both small and large presses have been developed which capitalize on the fact that some of the dampening solution is emulsified into the ink by literally running the dampening solution onto one of the ink rollers, and in effect eliminating the whole dampening-roller system. This innovation is still in its infancy, and there may be a long delay between its "development" and its introduction to the trade.

The many variables of the lithographic process are a disadvantage in comparison with letterpress because a relief press is not complicated by a third cylinder, a dampening system, or such an easily damaged plate. To these mechanical complications of the lithographic press must be added the chemical complications of the process. The two combine to create problems known in the trade as bleeding, chalking, crawling, blinding, scumming, and piling — the precise meanings of which need not detain us here. Were there a single product, many such problems could be eliminated by stand-

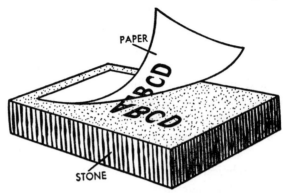

DIRECT: Original, or direct, lithography employs a thick stone block as a printing plate. The artist actually draws or transfers the image to be printed onto the stone. Separate stones are needed for each color. Flatbed presses are used and the paper is brought into direct contact with the inked stone. Once of great commercial importance, direct lithography has been largely replaced by offset lithography.

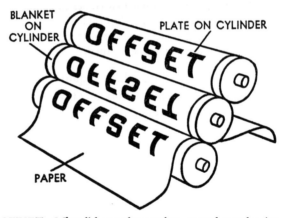

OFFSET: Offset lithography employs curved metal printing plates and a resilient offset or transfer blanket on cylinders. The image to be printed is usually "affixed" to the plate by photographic means. The plate bearing the image is printed on the transfer blanket and then offset to the paper from the transfer blanket. Separate plates are needed for each color and the work can be printed on multicolor presses or one color at a time on single-color presses. Offset presses are made for either sheet-fed (single sheets) or web-fed (paper on a reel) printing.

Fig. 4. The Lithographic Process
Drawings from Racine Lithographic Institute, Racine, Wisconsin

ardized procedures. But each press run is different, either in the paper or ink used, the location of the solids, or some other variable. For this reason much craft skill is required of an offset pressman, and he occupies a key role in the total lithographic process. At one time this role was shared more or less equally with the hand transferrer, and the earliest union in the industry was in fact an organization with membership in these two skills. The vacuum frame and photo-composing machine so completely replaced the hand transferrer that even the name is unfamiliar today, and, though other skills have gained in importance, the nature of the work has clearly left the pressman in a preeminent position. Yet an important qualification must be noted here. The relatively recent introduction of small presses has brought with it a very different kind of "pressman," whose skills are only slightly greater than those of a duplicating machine operator. Part of the rapid expansion of lithography in recent years has been possible precisely because press operation for certain types of printing has been so thoroughly simplified. As a recent advertisement asserted:

Your stenographers sit down while they type . . . why shouldn't a girl operate your offset machine from a sitting position too? NOW your short run offset reproductions can be run off . . . quickly . . . easily . . . almost effortlessly . . . by a relaxed operator, seated at the machine. Plates are changed in a jiffy . . . the blanket is automatically washed . . . clear, crisp, clean copies are delivered virtually into her hand . . . *all by pushing a few simple buttons . . . and everything* is done *from a sitting position!* [6]

There is no clear dividing line between these "offset duplicators," as they are sometimes called, and standard offset presses. One can buy a 9″ x 14″ (dimensions refer to the largest plate a press can utilize), suitable for running letter-size sheets (8-½″ x 11″). Between this and the 17″ x 22″, traditionally considered the smallest size press practical for commercial printing, there are a variety of sizes and types to suit anyone's taste and pocketbook. Several of them can be purchased for less than $1000, and the now flourishing second-hand market for such presses makes capital inputs almost infinitely divisible.

Although the first of these offset duplicators, the Multilith, was placed on the market in 1932, their major penetration has been since the introduction of simplified platemaking devices which contain all exposure equipment in a single unit and can be set to operate automatically. As in the case of the offset duplicators, their manufacturers claim great simplicity. There is "absolutely no attention necessary," they "completely eliminate all guesswork on the part of the operator," and of course "can be used . . . in any small office or plant." [7]

Simplified offset equipment has done much to spread lithography in the market for commercial printing; the web-fed offset press may well have the same effect in the publications printing market. Printing on a roll of paper rather than individual sheets avoids the problem of stopping a sheet just prior to its being gripped by the impression cylinder, and then being accelerated to the full speed of the rotating cylinder. Web presses can run at much higher speeds, printing two sides simultaneously is possible, and paper costs less because the paper manufacturer is not required to sheet and square the paper. But web presses are both expensive to buy and limited in their application since one dimension of the printed sheet is restricted by the circumference of the plate cylinder. For this reason they are unlikely to be introduced except in large plants, which have or are confident they can get the sort of work for which a web press is suited.

A Discussion of Innovations

Innovations have by definition changed the production process, but they also have played a central role in changing other elements in the economic context for collective bargaining. The impact of innovations is a major theme of this book, and it will be useful to provide a brief but general comment concerning their impact on the organization of the industry. Innovations have caused four significant changes.

1. *Lithographic firms are becoming less integrated, and there has been a rapid growth in the relative importance of the small firm.* New and more efficient production processes which have high minimum capital requirements will tend to increase the size of the optimum productive unit, and, so long as the extent of the market in which the firm sells is not a limiting factor, average firm size will increase. The large web-fed presses are exclusively oriented to products sold in the regional and national markets, but will be introduced only in those firms that can support the $500,000 to $1,000,000 outlay necessary to secure the technological economies available from this type of equipment. However, if the extent of some markets does enforce a limit on the optimum size firm, and the new process can be separated from the rest of the lithographic operation, then it may be introduced in a trade shop. For example, if ten small commercial lithographers could each use only one tenth of the productive capacity of a photo-composing machine, they could still keep a trade platemaker in business. The "productive unit" (the trade platemaker and the ten shops taken together)

is larger, but individual units are smaller, each of the ten shops having disposed of the operation which the photo-composing machine displaced.

The scanner, because of its high output and current complexity of operation, will probably be restricted to trade shops serving a regional or at the minimum a large metropolitan market. Most photo-typesetting machines can be installed for less than $50,000, yet unless a lithographer has a large and regular flow of orders requiring typesetting, these machines are, like the scanner, unlikely to be introduced except in the largest lithographic shops. Prior to the intrusion of lithography into product areas characterized by the extensive need for typesetting, this operation was performed typically in trade type composition houses. It is quite likely that this type of industry organization will continue, and the lithographic industry will adapt itself to it.

Should an innovation require heavier capital investment than even the largest firms can afford — for example, manufacturing presensitized plates — the operation may move out of the industry entirely and into supplier firms. The problem of producing plates of consistent operating characteristics is basically the problem of standardizing the plate-preparing and sensitizing operation. To do this economically is out of the question when the requirement is for only three or four plates per day, or even thirty or forty. Not even the largest plant could afford the cost of making presensitized plates, any more than they could afford the cost of making their own presses. Yet the value of reliable plates is such that presensitizing may soon extend to the longrun plate, thus eliminating the plate-graining and coating operations from all lithographic shops. On balance, the result of these changes will be a substantially lessened degree of integration and a decrease in the average size of firm.

Other innovations are not capital intensive and can be used most effectively by the small firm. For example, to adopt many of the contributions of the Lithographic Technical Foundation it is necessary to have certain skills and understanding but no great sum of money. This is also true in the case of simplified offset equipment. The cost of setting up a self-contained unit is remarkably low; $5000 to $7500 [8] wisely spent can put a letterpress shop or private plant into the offset business. This would include a small press, camera, platemaking, negative developing, and layout materials, with which equipment all run-of-the-mill printing could be produced.

2. *Innovations have eliminated some skills and required many new ones.* The much higher speed of the offset press as compared with the flat-bed press meant that more workers in the preparatory department were neces-

sary to keep the same number of pressmen busy. Flat-bed pressmen had to learn the mechanics of a cylinder press, and hand feeders saw their basic function being taken over by automatic feeders, several of which could be handled by the same feeder operator. Shortly after this trend began, the photo-composing machine and vacuum frame began the process of totally displacing the transferrer. The tendency for some platemaking to shift to the trade shops for commercial lithography speeded the elimination of hand transferring, and the fact that these machines were designed to make use of photographic negatives (or transparent positives) led to the more gradual displacement of engravers, designers, and poster and commercial artists. There was no need of transferring anything that could be photographed, engraving anything that could be photographed, or designing or drawing anything that could be replaced with an existing picture or something from which a picture could be made. Stone grinder, transferrer, poster artist, designer, engraver, flat-bed and rotary pressman, these and other job titles are largely history today, yet two generations ago such craftsmen dominated the industry. The membership records of the ALA provide interesting evidence of this change.

Table 4.   Union membership in four classifications, 1904, 1917, 1956.

	1904	1917	1956
Engravers	311	329	830
Transferrers	1021	1302	70
Flat-bed and rotary pressmen	949	753	330
Provers	133	137	260
Total journeymen and apprentice members reporting	2500	4782	26,500
Four crafts as per cent of total	96	52	6

Source: 1904: Lithographers' International Protective and Beneficial Association, *Proceedings* of the 1904 Convention (Philadelphia), President's Report, unnumbered; 1917: Amalgamated Lithographers of America, *Proceedings* of the 1917 Convention (Cincinnati), State of Association Reports, pp. 267ff.; 1956: ALA, *National Wage Survey, 1956*, internal record of the union.

Innovations have totally displaced some skills; they have also sharply reduced the demand for others. An innovation which has had this effect is the long-life plate. The series of changes which have led to the "indestructible plate" have altered the duties of the platemaker only slightly. What has changed radically is the number of platemakers required to keep the same number of presses running. In the 1955 convention of the ALA the problem is clearly expressed:

Delegate Gunderson: Now, in a town or city like Chicago, where we have a lot of big shops, we felt the impact of the bimetal plates. For example, in some of the large shops where they're using these plates and doing publishing work, they previously used about 35 sets of plates to run off these jobs that ran over a million runs. At the present time, they are using four plates and one additional black plate for changes in the type.

Now you can imagine what that does in a large plant.

Delegate Robinson: I know our can plants in San Francisco, where they have the tremendous runs of the can labels, the beer labels particularly, that formerly where they would make a dozen sets of plates to run off a given run of any make of can, now they call up the plate department and say, "Make me two sets," instead of 10, 12, or 14.

International Vice-President Mertz: I can speak for the general conditions covering this particular subject in the central region . . . the platemaking mechanics are the ones that are least called for in the entire group.[9]

Not only is there a loss of employment for platemaking craftsmen, there is a shift toward the increasing use of trade shops. The plant of a size that would justify investment in a photo-composing machine in 1947 would in 1960, because of the advent of the long-life plate, be better advised to make use of a trade shop supplier. Bearing in mind that the average-size firm has declined since 1947, it is apparent that the long-life plate will cause a change in the structure of the industry and of the workforce, the full impact of which is yet to be felt.

The innovations discussed above have totally displaced or reduced the need for craft skills, but they have required others in their place. The effect of the introduction of simplified offset equipment has been to turn the operation of this type of equipment at best into a semiskilled operation. There is only the most distant relation between a multicolor pressman and a Multilith operator, or between the operator of a photo-composing machine capable of handling 76-inch press plates and the operator of a plate printer. These men are not craftsmen in the conventional meaning of that word.

3. *The process is spreading to other industries.* The innovations introduced prior to 1930 did little to simplify the lithographic process, or make it suitable for competing with relief printing in any important product market. The introduction of simplified offset equipment since that time has done both. It is clear that the nature of an innovation can determine whether traditional users of the process or traditional suppliers of the market will be most likely to introduce it. The "direct image" plate is an example of how the nature of an innovation may determine which group of firms will capture (or retain) a market.[10] An image may be transferred to a direct-image

plate from a type form or electro by running the plate through a press as one would a sheet of paper. The facility with which copy made ready for printing by letterpress may be switched to offset makes this kind of plate and the equipment necessary to run it of great value to the letterpress printer and eliminates one of the important reasons for a letterpress printer not switching to offset. Reruns of old orders — a major item in a small commercial printer's sales volume — can often be made with greater economy by this method than by running them on letterpress equipment. Such innovations as these are oriented exclusively toward permitting a painless and profitable switch from letterpress to lithographic printing, and have made possible the phenomenal rise in the number of combination plants.

At almost the opposite end of the market, other technological developments have caused a rapid introduction of the process in the folding paper box and publishing industries. The web-fed press, for example, serves a market which until recently was closed to lithographers. Quite naturally not only lithographic firms will buy such equipment; publishers and large letterpress firms will adopt the process rather than have lithographers adopt their market. The web press is not efficient solely for products formerly in the letterpress domain; it also can replace sheet-fed lithographic equipment for certain work. For this reason the introduction of web presses not only provides a wider range of competitive suppliers for some former "letterpress only" products, but also for some former "lithographed only" products.

Additional competition is not a problem for lithographic employers alone, a fact which General Counsel Robinson has repeatedly emphasized to ALA convention delegates:

> What is the significance of these plants? The significance is that these plants are in direct competition for your jobs. Your employers simply cannot compete for this business without which you won't have any jobs unless these plants in their lithographic departments are organized upon comparable conditions.[11]

To many lithographers, within or outside the ALA, such comments are pointless, for the split in the industry is creating a split in the workforce, which is growing in importance with the increase of combination plants on the one hand and the growing number of large lithographic units in the folding paper box and publications printing field on the other. In general the lithographic workers in small commercial shops and combination plants require relatively less *craft* skill but a *wider* range of skills than do large-color or publications lithographers. Their employers compete in and are concerned with local competition, the large employers with national competition, and of course the identity of the competitors of each will be different.

In such circumstances these two groups of workers will favor conflicting organizing goals, jurisdictional strategy, and negotiating programs, and cooperation between them becomes difficult, particularly when the nature of reasons for this split is not clearly understood. When an ALA delegate to the 1959 convention arose to protest the policy of raiding organized shops, he couched his argument in terms appropriate to the worker in a small commercial shop. He was answered with references to the competitive menace of folding box plants, which for a commercial lithographer is no menace at all.[12]

4. *The elimination of differences between printing processes.* A number of recent developments may well have a more fundamental impact than any of the innovations discussed above, for these may lead to the elimination of process distinctions entirely. For example, there has been some limited success in eliminating the dampening system from the offset press and protecting the non-image area of the plate by altering the mixture of the ink. In relief printing, experiments have gone forward with "dry offset," which is basically printing from a raised image on an offset press. This process is now in commercial use and may become important if there is a shift to plastic plates. The plastic plate is a fundamental innovation that offers the possibility of making relief plates as simply and nearly as quickly as lithographic plates. Basically, the plate is a layer of plastic bonded to a metal sheet. When exposed to light, the plastic hardens throughout its entire depth, and the unexposed plastic may then be dissolved, leaving the light-hardened plastic at proper printing height. With such plates all preparatory operations up to the platemaking stage would be identical for both processes, and even the platemaking would only differ slightly. In fact, one type of plastic plate can be used in either letterpress or lithographic presses, and the choice need not be made until it is actually ready to put on the press. Such developments put identical processes under the jurisdiction of five different printing unions: the lithographers, the printing pressmen, the photo-engravers, the electrotypers, stereotypers, and the typographers.

5. *Innovations have changed the ALA and caused it to support technological change.* Innovations in products or methods are always at the heart of jurisdictional disputes, but they can also put other and less easily identified pressures on unions. Such was the craft consciousness of the lithographic pressmen that many years passed before the union could bring itself to cooperate with the feeders, and many more years before it would amalgamate with them. It is tempting to offer the introduction of the offset press as a direct pressure leading to the amalgamation of the lithographic craft

unions in between 1915 and 1918. There is no evidence, however, that the direct impact on the workforce was as important in explaining amalgamation as several other reasons less directly related to the introduction of the offset press. Though changes in technology had little to do with furthering the cause of amalgamation, they have certainly provided an important external pressure holding the union together. Taken individually, the demand for many lithographic skills has decreased precipitously; taken as a group, the demand has remained roughly constant or expanded. And, in the last twenty years, the displacement of men by machines has been far overbalanced by the demand for new skills elsewhere in the process. This growth has provided opportunities for the ALA; since 1939 its membership has trebled, while total union membership in the United States has doubled. But it has also provided problems. The rapid increase of lithographic workers has accentuated the decreasing homogeneity of the workforce, for many of the entrants have never heard of the ALA, or, for that matter, of "lithographing." They work in an "offset shop." Moreover, the workforce is relatively young. The average age of all ALA members in 1956 was thirty-eight; in only three small locals was the average age over forty-five.[13] People who joined the labor force after the 1930's test the value of unionism with a different set of memories than those who were breadwinners during that memorable decade. As ALA General Counsel Benjamin Robinson has described the situation:

> You face, first of all, what I have referred to before, the apathy of workers who have enjoyed full employment with reasonably good standards. They have the opinion that their skills alone have secured them their wages and conditions even in organized shops.
> And then what do we say in this over-all picture about our own union members who vote against their own union in N.L.R.B. elections? They say they don't want to antagonize the employers. They will pay union dues, but they don't want to be bothered.
> Oh, they believe in unions, they say, but they don't want the union to bargain for them.[14]

The very fact of technological change has created problems. To an individual worker there is little satisfaction in knowing that his unemployment is counterbalanced by the employment of workers elsewhere, and the lithographic union has reflected this feeling quite faithfully in the past and today in many local bargaining situations. In the last decade, however, an articulate and aggressive element in the national union leadership has sought to move the membership toward a wholehearted acceptance of technological change, believing that the wide range of skill classifications in the union can provide employment opportunities for all even if some of the classifications are declining in importance. The sustained high tempo of innovations and the accompanying expansion of the process have given impressive credentials to this group, and today the lithographers' union is known best outside the industry for its program of positive support for technological change. While a measure of this reputation must be attributed to good public relations counsel rather than actual accomplishment, the ALA has without question taken a more constructive stand on this issue than the typical American union. The following comments, for example, are keyed directly to membership, and not public attitudes. They were made by Local 1 (New York) President and International Councillor Edward Swayduck, in the course of a presentation by the Committee on Technological Developments to the 1959 ALA convention:

On organizing: What better argument can you give [nonunion men] than showing [them] you are fully aware of technological developments . . . pointing out to them that technological developments, if they are not harnessed, will cause unemployment? And a nonunion man hasn't got that kind of protection.

On a better life: In our union in the last twenty years, because of technological developments, we enjoyed three months of labor withheld, or vacations: . . . [ten paid holidays, three weeks' vacation, reduction from a 40- to a 35-hour week.]

On job security: We can never lose sight of the fact that we have to constantly watch out for allied trades because they too are developing technologically. If we do not wish to accelerate our technological developments and think small, narrow, then we will hold back technological development, and we shouldn't.

As they develop we have got to develop fast so that we are in a more favorable position competitively, because a mature union understands that only when his boss can properly compete that makes his job more secure.

There is little doubt that the lithographic union has lessened worker resistance to technological change. If technological change continues to expand job opportunities, the union will continue to support it.

Chapter IV

The Structure of Markets

The conditions under which employers sell goods or services have long been recognized as affecting arrangements in the labor market. A good deal of attention has been given to the extent of competition among firms so that interest has been diverted from other aspects of the market, less fundamental but still important in their effects. Competition sets the upper limits for returns to labor, but other features of the market, such as product variations, market mechanisms, and the locations of firms in relation to the market may also effect the terms of the labor bargain, as well as the structure of bargaining and even the intensity of interunion rivalries.

Three Dimensions of the Market

There is no single market for lithographed goods. The products range from beer cans to textbooks and are neither produced by the same firm nor usually purchased by the same buyers. The complex of markets can be simplified by viewing three of their aspects separately: (1) the type of product sold, (2) the process used (lithographic or letterpress), and (3) the geographic extent of the market. When boundaries for each of three market dimensions can be drawn sharply, as for example when buyers from Chelsea will accept no substitutes for handbills which are lithographed in a plant in Chelsea the market may be called insulated. The degree of insulation in each of the three dimensions varies over the full range of lithographed output, and is an important part of the setting for collective bargaining. For example, jurisdictional disputes are closely related to the number of markets in which both processes (letterpress and lithography) are used; that is, closely related to the number of markets which are not insulated with respect to processes. The scope of bargaining — national versus local — is related to the sharpness of geographic market boundaries. Similarly, wage relationships between firms are related to the insulation between specific types of product markets (that is, the degree to which one product can be substituted for another), and the wage relationships within a firm will be related to the number of these markets in which the firm sells. Each of these three market dimensions will be discussed in turn.

Type of product. Most lithographed products are unique. A lithographer can seldom produce for stock, and a buyer cannot resell his purchase. A million Esso road maps have little value except to the Esso Company. The product which is being sold, however, is the service of producing road maps to order, and this service is not unique. There are many potential buyers and sellers, the number of each depending on whether the product is "road maps to order," "road maps," or simply "maps." The Census description of "maps, atlases and globe covers" will widen the market definition still further, but at the expense of including firms interested in bidding on road map orders but not globe covers, or vice versa.

A firm able to supply atlases may also be interested in printing textbooks, as may a periodical printer who finds that advertising printers specializing in brochures are branching out into the periodical market. In short, there is a field of printing services so extensive that no one lithographer can attempt to cultivate the whole, and a set of circumstances so varied that no two lithographers will find the same part of the field exactly to their liking. Some of these differences are decisive, such as location in the metal decorating field, or the employment of skilled process-color artists in the prestige advertising field. Other differences may be so slight that even a trivial incident, such as the prospect of a bargain on a second-hand two-color press, may cause a shop to enter new product markets.

As firms enter these new markets, the new market mechanisms demand different responses and set a new scale of values on the conflicting goals of maximum speed, quality, economy and adherence to production schedules. For example, with the entry of a commercial lithographer into publications printing, the cost implications of a reduction in hours from forty to thirty-five may become less important, while the damage inflicted by a work stoppage will increase. These same cost implications are likely to move in the opposite direction for the commercial lithographer who expands into specialty printing. The market mechanisms and the market pressures on these two firms will be quite different, and, being so, will cause them to assess the cost of union demands quite differently. Yet more often than not they will be in the same bargaining unit if they are in the same city. Twin problems are presented here, one each for the employers and the union. The building of a united front by the employers is made more difficult, and whipsawing by the union facilitated. On the other hand, unless the union is willing to engage in a constant series of strikes and perhaps drive some employers out of business, it has to tailor its demands to the employers who can pay the least. The expansion in the number of markets which innovations have permitted the lithographic process to supply, quite apart from the question of jurisdiction or traditional employer allegiance, is tending to render the accepted city-wide pattern of bargaining unstable.

Process markets. The second dimension of lithographic markets is the degree to which a lithographed product and a similar product produced by other printing methods can be substituted for one another. Copies of the same text are almost perfect substitutes, however produced. Substitutability is less perfect as the distinctive advantages of each process become more important, as in the rendering of large solid-color areas (lithography) or sharp images (letterpress), or when a specific type of paper is required. Until about 1900 there was relatively complete insulation between the markets for lithographed and letterpress-produced items. No method other than lithography was suitable for producing multicolored prints, and that market was left to the lithographers. When "process color" was introduced to the letterpress industry — that is, the method of reproducing many colors by the use of only the three primaries — complete insulation was lost. But the three-color process did not penetrate seriously into the lithographic market until 1906–1907, when the long strike at that time forced many buyers of color-work to look elsewhere for suppliers.

While the three-color process was blurring the clear market boundaries between the two processes in the field of color printing, the offset press was doing the same in the field of commercial printing. With greater speeds and lower plate costs made possible when art work (or later photography) had to be combined with a small amount of text matter, the offset press permitted lithographers to undercut users of equivalent letterpress equipment and enter some new markets. Improvements which simplified the process and reduced capital costs gradually increased the overlap of printing processes in the commercial printing market. The introduction of small offset presses such as the Multilith was the most notable of such developments. The impact of these innovations is reflected in this comment by a letterpress printer in 1957:

Recently we had about fifteen forms that required less than five hundred copies each. In the past we had used our letterpress equipment for all form work. This time we gave it to a small offset printer whose running and paper bill was only $26.00. Typing of the offset masters cost another $15.00 or a total cost of $41.00. Due to the ruling on some of the forms, we estimated that typesetting alone would have cost over ninety dollars. Setup and running of 15 forms would have cost us about thirty dollars. Although the quality would have been better if we had printed the job on our letterpress equipment, the savings of almost 70% gave us no rational alternative.[1]

Another innovation which introduced a substantial overlap between process markets was the long-life plate. For example, "textbooks" is a moderately narrow product class, and one which formerly was completely in the domain of letterpress. Increased plate life put some of the shorter runs of illustrated texts within the range of lithography, and as plate life increased further, so did the overlap between the processes. This overlap continues to grow with the introduction of the web-fed press to lithography since with this equipment the high printing speeds possible with continuous feeding are available to both processes. The cost differential between processes here is slight, the advantage being determined by minor differences in order size, paper requirements or type of copy, differences which may easily be less important than the efficiency of the printing firm. Photographic typesetting equipment has had a similar effect in breaking down the insulation between process markets; with its introduction one can no longer assume that straight text matter can be done most cheaply by letterpress.

In the long run all these overlaps would be eliminated as each process claimed the market for which it was the low-cost method. But for reasons to be noted shortly, this matter of sorting out markets is time-consuming, increasing the jurisdictional conflicts between printing unions gradually, as it permits them to marshall their forces for a potentially long and bitter struggle.

The final dimension concerns the geographic extent of markets. Display material for nationally-advertised products may be bid on by firms in New York, Chicago and San Francisco. The small independent grocers and druggists of a suburban shopping center, on the other hand, will seldom go to the other side of town to buy their handbills or mail advertising. Between these two extremes of the national and neighborhood markets there are infinite gradations, but for our purposes only two types need be distinguished, the national market and the local (or metropolitan) market.

In lithography the basic determinant of firm size is the extent of the market available to the firm. If a firm wishes to compete for national advertising or the printing of a national publication, it must grow to (and may perhaps go broke on) a far more impressive scale than the lithographer serving a local market. For example, one of the commercial lithographers interviewed in Boston whose sales were between $1.5 and $3 million, needed a four-color 76-inch press to break into the color lithography and publications market. The cost of the press was $300,000, or from 10 to 20 per cent of current annual sales. Yet his first bid on a major order was rejected because the publisher was not willing to risk placing business with a company with only one such press in case of breakdown.[2]

On the other hand, it is doubtful if there are any large firms (over 250 employees) that can compete for the bulk of local printing work. There are a number of reasons for this, and in order to discuss them it will be necessary to anticipate some of the argument presented later. It is of course true that large firms avoid small orders because they make less money on them. Among six large firms interviewed, five had a minimum order size varying from 5000 to 15,000 impressions, and the sixth subcontracted this work to other shops.[3] But small firms are not noticeably less profitable, and so it seems clear that small orders are not less profitable *per se*.

One important fact is the different technological requirements for handling small orders efficiently. One of the distinctive attributes of the local market is that it may be serviced adequately with capital and labor inputs which can be varied widely in proportion to each other and in absolute amount. Both trade skills and capital investment can be reduced by using premixed chemicals, inks and presensitized plates. This economy is available to firms of all sizes but is most relevant to the type of product sold by the small firm. It has been said of lithography as of the other printing trades that the main thing needed to go into business is dissatisfaction with one's present job. A person of no great experience, but with five to ten thousand dollars[4] can buy equipment which will expose and develop presensitized plates with a minimum of operator skill and small presses which a competent office employee can operate. Experience can reduce capital requirements by making older, nonautomatic machinery feasible, or broaden the market available to the new entrepreneur through his ability to operate larger and more complex equipment.

No less important are the different administrative requirements. It makes no sense to route an order through costing, planning, layout, materials requisition, and so forth, when the total assignment is exposing the plate, slipping it on a Multilith and running off 2000 handbills. Finally, the market requirements are different. The man who can get the order for printing American Airlines timetables is not the man to call on the neighborhood restaurant for his menu business, nor will his selling strategy be remotely similar. The desirable skills, knowledge, and the quality of the personal relationship between the buyer and supplier are quite different, so also is the relative importance which the buyer attaches to cost, quality and prompt delivery. The owners of some large firms have attempted to straddle these problems by setting up independent units, but such a practice is not common.[5] It is safe to conclude that, just as participation in the national market enforces some minimum size limit on the firm, participation in the local market enforces a maximum size limit.

It is worth repeating that there are no clear dividing lines between product markets, between process markets, and between the different geographic markets. Each of the three market dimensions represents a continuum along which convenient points or subdivisions have been treated as separable and distinct elements of the market, but which of course they are not. Forbes Lithograph does 75 per cent of its business with customers outside of New England, but will compete on some orders with Spaulding-Moss and Buck Printing who do 75 per cent of their business within the region. These two companies will in turn compete with firms who serve only the Boston metropolitan market, and so on. Similarly, the distinction between color and commercial printing, and the distinction between products in which there is process competition and those in which there is none are abstractions that make discussion of the market possible, but still remain abstractions.

The infinite gradations in each of these dimensions are themselves matters of interest because they make any strategic decision concerning jurisdiction by the union or an employers' association immensely difficult. On the one hand, neither organization can hope to secure uniform supplier policy (or workforce control) in all product markets, yet any coverage short of this will be made partially ineffective by firms outside the organization which are competing in the market.

This competition is not "pure" in any of the markets, nor the same in any of them. In order to assess the impact of competition in the product market on decisions made in the labor market, it is useful to look more closely at these different market relationships, and in particular at the market mechanisms which implement them.

Market Mechanisms

The methods of determining prices in the lithographic industry vary from straight price-quoting by the lithographer to cost-plus arrangements determined by the buyer. In describing these mechanisms and ascribing reasons for their existence, the industry view will first be considered.

Costing methods. Methods of assigning costs to output vary greatly among firms since a competent cost accountant is a luxury which few small firms can afford. One of the most significant services of the three industry trade associations is the provision of costing systems for their members. Equally valuable is a related annual ratio study of the Printing Industry of America, which compiles a large number of accounting ratios from statistics supplied by several hundred printing firms, thus providing a norm against which individual firms may judge the results of their own financial management. These services presumably lessen the variations in product costing between firms, but they by no means eliminate them. It is not uncommon to have a large spread between high and low bids when sealed bids are used to choose suppliers. Lithographers are never more eloquent than when recounting instances of other firms bidding less than the actual wage costs of a job. The cost accountant of the National Association of Photo-Lithographers reported recently that of seventy firms visited, only twelve knew their actual costs. His comments on one of the plants visited is revealing:

> When I appeared in this plant ready for work, I was informed that they had a complete up-to-date cost system in effect.
> I learned that they were working a full two shifts but they were not making money. I found their spoilage was heavy, but spoilage records were not maintained . . . And finally, after breaking bread and really getting close to the estimator, he confided in me that he was worried, because the firm was only getting 80% of the jobs that he estimated. I have always thought 30% was good.
> When I had completed my presentation of this firm's cost rates, the president sat back and said, "Frank, we would not get 25% of our jobs if we used the rates that you have come up with." [6]

Markup is the most common method for setting prices, but price lists compiled for standard items may also be used, and individual job prices set from these. The price lists themselves may be built up from budgeted hourly rates, or, on occasion, from competitors' price lists. Perhaps the most direct pricing policy is simply to multiply labor costs times a constant, 2, 2.5, 3, and so on, thus saving much paper work, and driving competitors crazy. A successful pricing policy has been said to depend on a genuine costing system, budgeted hourly rates for each cost center, and a good estimator.[7] Very few small firms have even two of the three requirements.

Problems in defining the product. Some part of variations in bids received by a buyer is attributable to different interpretations of what constitutes fulfillment of his order. It is in practice impossible to establish quality standards in advance of production, and a buyer is obliged to balance quotations received against his knowledge of the bidder's reputation for quality and speed of delivery.

A moralistic little tale in one of the trade journals illustrates this point:

> A supply company recently got an assortment of bids on a job for a new catalog. Among the bids was one which was a third less than the next lowest. Somewhat against his better judgment, the customer accepted this bid.
>
> Then the fun began. The printer kicked at the art work and everything else he could kick at. The customer refused to change his order in any way. The printer was slow about delivery; the customer refused to offer premiums for speeding up.
>
> Delivery was eventually made and the customer refused it, saying that it was sub-standard work. Neither side would back up and the deal stalled.[8]

In the first place there is the very real problem of establishing a usable definition of "product." A recent survey by one of the trade publications found it necessary to specify nine elements (quantity, paper stock, delivery date, and so on) for each of three items on which price information was requested.[9] Multiply this by the large variety of products, defined with sufficient precision to permit pricing, and the mere size of the resulting price list would discourage the hardiest association executive.

The effort of the NRA Lithographic Code Authority to enforce fair pricing was a failure, and not before or since has anything more than uniform costing systems been attempted for the full range of lithographed products. An effort in 1888 by Secretary Koerner of the newly formed employers' association "to ascertain the exact cost of lithographic printing, without entering into the realm of theory," ended in failure:

> It may not be placing the intelligence of the lithographer on a very high plane, by making the statement that the wide fluctuation in prices, the curious but positively disastrous methods for securing work, and the frantic efforts to turn it out for the price when secured, is due to a great extent to the absolute barrenness of data and the absence of a single intelligent effort to demonstrate what the cost of production really is.[10]

Buyer participation in costing. There has been no noticeable improvement in pricing practices in the seventy years following Secretary Koerner's initial investigation. Under these circumstances, it is not remarkable that price quoting is an unacceptable method of price determination for many types of printing purchases. The larger the order grows, both in absolute size and in relation to the total expenditures of the purchaser, the greater will be the concern for differences in price between a number of bidders, and the greater the interest in assuring that the lowest price has in fact been secured. The conventional method of receiving this assurance is to secure a number of bids, trusting in the competitive mechanism of the free enterprise system to provide the desired quality at the lowest price. In companies of sufficient size to have the purchasing function vested in a person other than the manager, this faith is formalized in the requirement that, in all orders above a certain sum, competitive bids must be secured.[11]

We have already noted that there is sometimes great difficulty in establishing measurable quality standards, and in such cases the lithographer is advantageously situated in that he can sell according to a more complex and less clearly defined set of values than price alone. A large color lithographer quoted a price to a regular customer on a job on which three other bids were also secured, all of them being lower than the interviewee's. When called by the customer, he agreed to shade his price about 10 per cent, but when the order was delivered he received a complaint on the quality. His response was prompt: "Sure, I know that isn't ——————— quality, but if you want me to compete with prices from second-rate firms, I'll have to compete with them on quality too." [12]

It is more commonly the case that on large orders a customer knows exactly what he wants and will not permit even small variations in price to go unchallenged. In such cases the customer cannot safely give an order to the lowest bidder without knowing the firm's reputation for quality. The more able the buying firm is in specifying precisely what is required, the more necessary it is to know which sellers can meet these specifications. For this reason the number of firms which are effective competitors may be so limited that the buyer, either because he mistrusts the independence of the bids received or the ability of the supplier firms to estimate with accuracy, may request bids on the basis of a detailed cost breakdown, for producing the order. On occasion the buyer may enter directly into the scheduling of production and decisions concerning the materials or methods to be used. It is a common practice for buyers to arrange the purchase of the paper to be used, particularly when the same order is being produced by firms in different parts of the country to supply the requirements of their respective regions. Quite naturally lithographers are reluctant to have buyers enter into their costing process, and buyers are permitted to do so only when the lithographer is compensated in some manner for the concession. Sometimes this compensation may be merely the chance to remain in business, but more often it is the assurance of a stable market. Accompanying such supplier participation in the internal pricing process there is normally a supply contract, an agreement which states in essence that the buyer-supplier relationship will continue as long as prices will not vary for reasons other than reasonable changes in factor costs.

The supply contract. The supply contract has been a standard device in the publications industry for some time and has been used in a limited group of lithographed product markets since at least the 1920's. Its importance lies in the extent to which it has spread to a significant proportion of the lithographic industry as firms have gone into the publications printing field. Supply contracts are of importance to the printing buyer when regularity of supply is a basic part of the service being purchased, as in the case of timetables, directories, private house organs or trade publications. The cardinal sin of the supplier is neither overpricing, since the cost build-up has been established, nor poor quality, because standards there also have been previously determined, but an interruption in supply. On one-time orders a delay in delivery may be cause for serious complaint, and future business from the customer may be lost. But when a firm is operating under a supply contract, a delay in delivery may result in its losing a substantial, continuing piece of business, around which the production operation has been planned. This, combined with the fact that competition for such business is not only with other lithographers but also with letterpress firms which formerly had the field to themselves, places the firm faced with a potential work stoppage in a peculiarly difficult position.

A somewhat similar problem faces the firm which has installed its own lithographic department, as in the case of the can companies and folding paper box plants. In such operations the principle of the supply contract operates with particular force, for the productive operation has been built around the assumption that a single supplier, in this case a division or department of the company, will be providing a continuous flow of the lithographed product to the rest of the operation. The technological pressures which favor the integration of such operations with the rest of the production process have been described in the previous chapter. The economies which the integration provides, however, are heavily dependent on an uninterrupted operation.

Trade advantages. The market mechanisms described above apply in the main only to products sold in the national or regional markets, which is to say, only to "large" firms in the terms of the industry. In the neighborhood or local market a different group of products is sold under a set of market mechanisms quite different from those described above. Two important differences — less marketing experience among lithographers and less understanding of the printing market among buyers — explain the importance of nonprice competition in this market.

It is nearly impossible for a lithographer to differentiate his product. Lithographers, so important in helping their customer differentiate products by advertising and distinctive packaging, have no way of branding their own stock in trade. One way of getting out of this perennial gale of competition is to emphasize not the product but the service, and to offer advantages to customers which make the individual lithographer's service unique. Some of these advantages are standard among large firms, and others dwindle to insignificance when offered to a customer who thoroughly understands the printing market. But among small firms, the lithographer who offers them may well be able to keep his prices firm in the face of underbidding by his competitors.

Service: Why yes, it will cost a little to change the caption, but or quotation stands. We're here to serve you, not haggle over pennies.

Delivery: That's right, and we'll stand by it. We'll take a loss rather than make a late delivery.

Quality: Sure, you can get it cheaper. There's two or three firms that will produce that kind of work for you.

Proximity: By all means, John, drop over any time. I expect we'll plate up around 11:00 A.M.

Customer Concern: That's really hard-hitting copy you sent over for the handbill . . . That part where you tell them you sell the cheapest suits in town is another one. I wonder if you'd like us to just rephrase those a little?

These trade advantages are not so much added to a product as they are offered to a customer. This is not a trivial difference. In the local market, the basic unit of significance is not the demand of many customers for a single product, but the demand of a single customer for many products. The small printing buyer rarely shops around for the best buy on calling cards, again for handbills, again for direct mail advertising, and so on. It is much too time-consuming to spell out specifications in detail and to repeat the process several times whenever additional printing is required. Moreover, each time a new supplier is engaged, there is a whole new teaching-learning process to be gone through. Does "You'll have it tomorrow morning" really mean tomorrow morning, or sometime the day after if the paper comes in? Common understandings of the meaning of such time-saving phrases as "There's no hurry on this job," "It doesn't need to be quite so plush as last year," can save much time and out-of-pocket costs, just as when the requirement of proofs can safely be waived. Quite naturally the supplier views this attitude of the buyer with some equanimity. It is pleasant to know that, even if the delivered price on an order has been set with great moderation, it will not be compared with some outrageously low figure set by a fool who does not know his own costs.

House accounts. A supplier will nearly always be willing and eager to handle all the printing requirements of a customer, although in some of them his costs may be greater than a competitor's. His price on such work will typically be related more closely to an "expected" price than to his cost of production; it is unreasonable for him to open himself to the charge of gouging when the lesser return can be made up on other, more profitable orders. The ideal situation for the commercial printer is to have customers so firmly attached to his firm that comparison of his prices by securing competitive quotations would seem an act of bad faith. Such customers are called house accounts, and it is perfectly possible that the best indicator of a firm's profitability is the percentage of sales volume coming from such accounts.

As already indicated, this buyer-supplier bond which converts a customer into a house account does not have a purely or even primarily economic explanation. It is true that both buyer and supplier would dispute this. In the words of a trade association manager, "The reputable lithographer would contend that his [house] accounts fared better economically than open-bidding buyers, and in all likelihood the executives of the customer would agree." [13] It would be foolish for the seller to contend otherwise, and remarkable if the customer did not agree with him. Few buyers would admit that such decisions were not based on a dollars-and-cents logic. Yet the fact remains that the customer's belief that he fares better economically is based on faith rather than competitive quotations. Regardless of formal explanations, the strength of the buyer-supplier bond depends heavily on the degree to which the customer is interested in convenience and avoiding decisions more than in minimizing cost, and will permit personal relationships to affect what are conventionally considered economic conclusions.

The element of convenience is in certain respects best seen as the opportunity to pass decision-making on to the supplier. Numerous printers attempt to facilitate this by urging customers to give them the end use requirements and then to leave the burden of deciding what sort of printing will best accomplish this to the printer. There is much wisdom in such an approach:

Take a whole problem of a businessman, and you can get out of price competition. Give him complete service on an advertising problem, find out what he wants, keep in touch with him of course to see that he likes what you're doing, and you can make some money. The advertising boys are beginning to do some of their own printing, so the thing to do is to go out and get your own advertising business.[14]

Such an approach is less effective when the buyer is a large firm and the order is substantial. The market machinery adapts itself to the different situation, and seller strategies have to be altered accordingly. Compare the following excerpt with the interview note given above:

> So much money was spent [by lithographers] on speculative art work that the advertisers discouraged this form of sales expense. The advertising profession as a whole was maturing . . . and realized that advertising literature and dealer material was part of an over-all campaign to move goods to the consumer, and was not competing with publications as advertising media . . . Gradually lithographic sales changed from trying to solve advertising problems to solving printing problems. The agencies, in turn, stopped trying to collect a commission from the printer for printing orders placed by them, but charged the client for the work done on this part of the campaign.[15]

"Creative selling" still plays a role in building national house accounts, but its importance is declining. In local markets, however, the buyer is often a small businessman who is his own purchasing agent and sales manager, with complete authority to turn over his advertising, packaging, and other printing problems to his supplier for proposed solutions. Quite naturally, the businessman's willingness to get rid of problems foreign to his basic field of competence will be increased if there is a satisfying personal relationship with the supplier. Perhaps for this reason, house accounts are most common where the extent of the market permits easy face-to-face relations between the buyer and the supplier.

The buyer-supplier bond. The connection between the lithographer and his customer is an important part of the market structure, and it is useful to look at it briefly in more abstract terms. A typical demand function for an uncommitted buyer is shown in Diagram A.

The demand curve in Diagram A may be thought of as a schedule of the amounts of printed materials which would be bought in relation to a corresponding series of prices. It shows that for the large bulk of printing buying, the product is not elastic (price sensitive), and that decreasing the price over much of the curve will not enlarge the amount bought by a single purchaser. It might, however (below point R), cause the buyer to switch some mimeographed work to Multilith, or perhaps add to his direct mailing campaign.

From a seller's point of view, the demand of any uncommitted buyer is

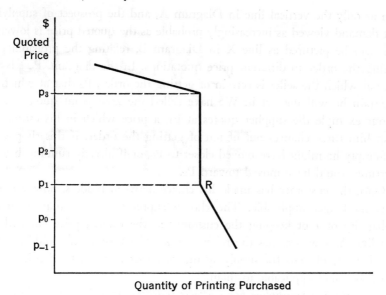

Fig. 5. Demand of a Single Buyer for Printing

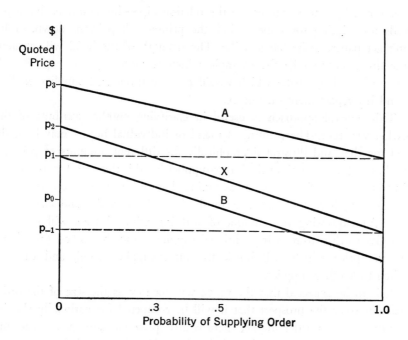

Fig. 6. Supplier's View

seen as only the vertical line in Diagram A, and the prospect of supplying that demand viewed as increasingly probable as the quoted price is lowered. This can be pictured as line X in Diagram B, relating the probability of getting the order to different price quotations. In the diagram, P_{-1} is the price at which the seller is certain of getting the order, P_2 that at which he is certain he will not get it. We have called the zero profit quotation P_0. In our example the supplier quotes at P_1, a price which in his estimation gives him three chances out of ten of getting the order. If this shop were half-empty he might have moved closer to P_0, or if already running heavily overtime would have moved toward P_2.

Once the customer has made his choice between bidders, however, line X is no longer applicable. The chosen supplier assumes that he has a probability of 1 of keeping the customer at the quoted price P_1, and for him line A now becomes the relevant one. The unsuccessful bidders have a probability of zero (of supplying the customer at the quoted price), and line B becomes applicable to them.[16]

How the buyer-supplier bond affects industry organization. It should be noted that the vertical distance between lines A and B is founded on reasons that will continue to be valid if the actual buyer-supplier relationship is interrupted by causes external to the relationship — for example, the buyer needs no printing for some period, the printer's shop burns down, or he can't get paper, or he has a strike. The strength of the bond will weaken over time, but to take the example which is most relevant here, it will have to be a long strike which would put the traditional supplier on line B and his replacement on line A.

This general approach is useful in clarifying another attribute of the market structure. The printing demand of individual buyers may be gathered together and illustrated graphically for different amounts of printing purchases. The strength of the buyer-supplier bond is then most conveniently illustrated, as in Diagram C, by assuming that all buyers are connected with specific suppliers, and that as a result there are two demand curves, the first (A) expressing the amount of printing which buyers will purchase at a given price from their traditional suppliers, and the second (B) showing the price which will break the buyer-supplier bond, and cause a switch to another supplier.

It is to be expected that the range will narrow as the size of the order increases since the prospect that it will be competed for nationally also increases. In this broader market there is little price latitude open to the supplier. For reasons to be explained shortly, at the upper end (smaller order

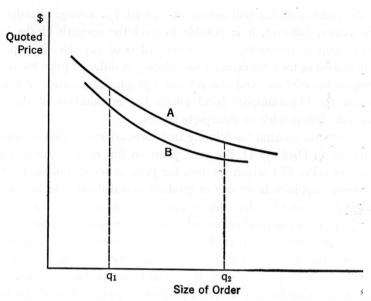

Fig. 7. Strength of the Buyer-Supplier Bond at Different Order Sizes

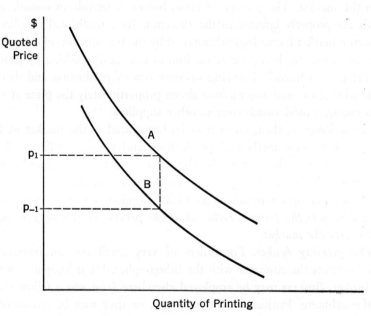

Fig. 8. Demand for Printing in a Single Market

size) the printing broker will narrow the spread. For a range of order sizes in the center, however, it is possible to treat the strength of the buyer-supplier bond as reasonably constant for orders of any size. This permits the illustration of the total demand for printing at different price levels without respect to order size and also permits a graphic explanation of the way in which the buyer-supplier bond affects the introduction of the offset process into lithographic or letterpress houses.

The shortrun demand for printing in a metropolitan market is expressed by curve A in Diagram D. For each point on this curve, there is a point on another curve (B) which specifies the price at which this demand will shift to new suppliers. If the cost of production is defined to include normal profit, and it is recalled that any supplier sees quantity as given and price the variable, then the equilibrium price-quantity will be the point at which the cost of production of the traditional supplier group intersects with curve A at price P_1. In order to shake the business loose in the shortrun, outsiders will have to quote at price P_{-1}, well below production costs.

If, on the other hand, a different production process is used that lowers the cost of production to P_{-1} or below, then the market can be claimed, and a new set of traditional buyer-supplier relationships forged which would with the passage of time exclude ever more effectively the former suppliers from the market. The passage of time, however, introduces considerations which are properly ignored in the shortrun. As a traditional supplier sees his secure market increasingly threatened by the introduction of a lower cost production process, it is possible for him to protect his cushion by introducing the process himself, lowering his own cost of production and therefore the market price, and also pushing down proportionately the price at which his accounts would switch over to other suppliers.

In the longrun, then, offset may be introduced to the market as it becomes a lower-cost method of production, and the cost differential need not be as great as that implied by the strength of the buyer-supplier bond. But firms which are already in the market, that is, the letterpress firms, will enjoy a comparative advantage in its adoption. *The market will determine who gets the process, rather than the process determining which industry gets the market.*

The printing broker. For orders of very small size an intermediary often connects the customer with the lithographer. He is known as a printing broker. Brokers may be employed elsewhere (one was a fellow student at Massachusetts Institute of Technology), or they may be employed full time. It was noted earlier that the management of small firms most often is drawn from men with production experience, rather than selling, accounting, or administrative backgrounds. The printing broker is really a free-lance commission salesman, who, at a price, performs the selling function for the printer. He is a large buyer, but of usually small order sizes, who is price-conscious to a degree common only among much larger printing buyers. The permissible spread between production cost and sale price thus continues to rise, the smaller the order, but for many of these orders the printing broker is the one who slices off most of it.

On occasion, a printing broker may secure a series of house accounts which would be considered desirable customers for a fair-sized commercial lithographer. Here, of course, the percentage take of the broker is less than that for, say, personalized stationery, and the security of the account is endangered by the prospect that some lithographer, perhaps his own supplier, will convince the customer that cutting out a middleman will have a favorable effect on the price. This is by no means an idle worry for printing brokers or for any firm that takes an order that must be let out to another firm for completion. For in one sense many small and a few large firms are printing brokers, in that orders are accepted, either for the prospect of a profit or just to keep the customer happy, for work that is either not possible or practical to do within the firm. In every large city there are firms that capitalize on this understandable mistrust of suppliers by setting up as printers to the trade. In Boston, for example, Hub Offset is the outstanding firm of this type, and each of its excellent advertising inserts in the trade publications ends with the phrase "No salesman will call. We have none." Also advertised is its "Jiffy Estimator," a device described to me in one interview as taking all the work but none of the profit out of printing. A portion of the introduction to this booklet is given below:

TRADE PRICE LIST OF OFFSET PRINTING

This pocket estimator and price list is designed to help you figure the most complicated offset jobs quickly. These are trade prices offered to the trade only. The Hub Offset Company does not solicit or accept work from the end user. Your customer is always *your* customer . . .

Hub's prices are designed for you to make a profit. Add 25% to get normal retail prices.[17]

The printing broker is another of the market mechanisms that makes the small firm a viable economic unit. Just as the trade shop helps it surmount the lumpiness of capital requirements, and the trade association does the same for managerial skills, so the printing broker provides a way for small firms to reach markets which are otherwise barred to them from their own lack of marketing skills.

The Impact of the Market on the Lithographic Industrial Relations System

As the lithographic market has increased in extent and variety, there has been a loss of employer homogeneity, and with it a loss of employer unity at the bargaining table. Much of the market expansion has been in products having a local or metropolitan market, which has strengthened city-wide bargaining. Yet the national market remains important, presenting serious problems to employers in high-wage areas who serve this market, and to the union. The employers serving the local market are faced with a quite different set of competitive problems, yet a single city-wide bargain must be struck if all lithographers are to be covered by the same set of working conditions.

The different types of demands, such as wage increases, extra overtime premiums, shorter hours, longer holidays, restrictions on use of apprentices, on the manning of equipment, on flexibility in the use of craftsmen, all can be given a money value in the managerial mind. But the values given will vary widely among different firms. It is also true that the pressures which the union may use to enforce its demands will be viewed differently. For example, color lithographers and trade shops require highly skilled craftsmen, perhaps more than in publications lithography and certainly more than in commercial lithography. For this reason the intrinsic bargaining power of individual craftsmen is greater in color houses and trade shops. Yet the economic damage of a strike is less than would be suffered by a publications printer, whose operations are built around a small number of large contracts to supply a continuing flow of printing. These contracts change hands rarely. Commercial lithographers, because of the buyer-supplier bond, are in general most able to resist, and often the most likely to, because the existence of bedroom printers and unorganized shops sensitizes them to the existence of low-wage competition, which by and large the color and publications lithographers do not face.

Of fundamental importance is the steady elimination of distinctive market boundaries between letterpress and lithographic firms. A jurisdictional dispute between the ALA and the Printing Pressmen's Union (IPP&AU), begun when the latter feared that process competition from the offset press would cut into relief printing markets, lay dormant for thirty years until such competition became a fact. The rapid expansion of suppliers for lithographic markets has presented so serious a challenge to the lithographic union that it stood in readiness to leave the AFL-CIO almost from the time it entered, and immediately did so when ordered to give up its right to "organize all lithographers, wherever they may be."

The right to organize all lithographers is meaningless without the ability to do so, and the expansion of the lithographic process into the market for commercial printing has put the ALA under a severe organizing disadvantage. For in this market the buyer-supplier bond is very strong, and leads to the introduction of the process in firms which often have a secure connection with other unions. The challenge that this presents to the ALA is measured not only by the extent to which the products produced in these firms compete with the products of ALA-organized firms, but also by the extent to which other workers are being equipped with skills identical to those held by members of the ALA. The expansion of the lithographic market has unquestionably strengthened other unions in the industry, notably the IPP&AU.

Chapter **V** ⑤

The Workforce

In view of the wide dispersion of the lithographic process, one can expect that accurate estimates of the size and distribution of the lithographic workforce will be difficult to provide. Even government offices, which use identical definitions for "employee" and "industry," cannot agree.[1] For example, the Bureau of Labor Statistics thought there were 60,500 employees in the lithographic industry in 1954, but the Bureau of the Census placed the figure at 77,700, or 28 per cent higher. For 1958 the Census figure of 97,500 is 48 per cent higher than the BLS figure. There is an additional problem. Interest here is not in all employees, but only in those who are members of or competitors with the organized workforce.

These three problems — identifying lithographic workers outside the lithographic industry, choosing between conflicting estimates of the number of workers in the industry, and deciding which groups of workers should be excluded from our definition of the workforce — form the major subject matter of Appendix B. Some of the methods used in this Appendix are scarcely better than flipping a coin, but it has seemed important to provide some estimate of total workforce, however imprecise, and the results are summarized in the list below:

Lithographic industry	58,700
Metal decorating	3,500
Captive paper plants	7,000
Private plants	3,000
Combination plants	15,500
Publishing houses	17,600
	105,300

In any attempt to describe the skills distribution within this group of about 105,000 workers, one must turn almost entirely to nongovernmental sources for useful information. The method here has been to use the distribution of skills within the ALA as the basic estimator and to modify this when evidence suggests that their degree of organization is not the same in all categories of workers. It is clear for example, that the ALA has organized a larger fraction of four-color pressmen than one-color pressmen, and that their degree of organization among offset duplicator operators is very small. Again, the odd bits of information on which an estimate must be based are gathered in Appendix B, and only the resulting estimates are provided here:

Preparatory trades	15,000
Platemakers	5,500
Multicolor pressmen	4,000
One-color pressmen	11,000
All other pressroom crafts	10,000
All other lithographic journeymen and apprentices	3,500
Total journeymen and apprentices	49,000
Other lithographic production workers	56,000
Total production workers	105,000

The Lithographic Job Clusters

The skills distribution within the workforce of an industry is an important determinant of the collective bargaining structure of the industry. While the nature of skill requirements and proportion of the workforce requiring craft skills is determined by the production process, the distribution of skills is more directly related to product market factors. For example, a firm specializing in advertising brochures and counter displays will need more color cameramen than a publications printer, and both will require more highly skilled workers than a lithographer serving a local market for commercial printing. In each type of firm, clusters of related jobs will be found, different for different types of firms but remarkably uniform among firms serving the same product market.

The phrase "job cluster," first used by John T. Dunlop and Robert Livernash, identifies the commonly noted closeness of wage relations within groups of jobs in a plant.[2] Thus the wage relations of toolroom jobs are closer to each other than to any clerical jobs, clerical jobs are in closer relationship to each other than to any unskilled factory jobs, and so forth. In addition to the similarity in duties, the administrative organization of the production process will help determine the content of job clusters. For example, one would expect one set of job clusters to form if subcontracting were the rule in any industry, another if all firms were integrated shops, each with its own policies on layoff, transfer, promotion, and so forth. A final element that links jobs together is social custom. The support for such custom may arise from within the plant, from the community, or from the way workers are organized in trade unions.

Each of these three links — technology of the workplace, administrative organization, and social custom — plays its role in shaping the lithographic job clusters. Pressures from technology and the nature of union organization tend to reinforce each other, and thus appear most important, but one need only consider the increase of lithographic workmen in the IPP&AU as a result of the rise of the combination plant, to recognize the vital importance of how the production process is organized.

In any job cluster there are one or more "key" jobs, on which attention is focused when wage or other changes are being considered, and which set the pattern for all other jobs in the cluster. Thus it is possible to make interplant or even interindustry comparisons of wage changes by the use of a relatively small number of rates, those for the key jobs in each of the relevant job clusters.

There are two job clusters in firms producing color and publications lithography. The cluster of most highly skilled jobs includes two groups, the first being the preparatory crafts in which the three skills of color cameraman, dot etcher/process artist, and stripper are the key jobs. In the pressroom is a separate job group with the multicolor pressman as the key classification. Most often these two job groups are so closely connected as to form a single job cluster. The rest of the production workers in this type of firm, roughly 40 per cent of the total workforce, are not a part of this high-skill cluster. No investigation has been made of the relationship this latter job group has to similar jobs in the local market, or to jobs in similar firms on a broader geographical basis. It is of interest, as we shall note in a later chapter, primarily because the ALA decision to organize such workers in the late 1930's has come back to plague them two decades later.

In trade shops there is a single cluster of lithographic craftsmen in which the key jobs are again the three preparatory crafts, but not including the multicolor pressman, a kind of skill which will only rarely be found in such plants. This cluster is in job duties similar to its counterpart cluster in color houses, but the product market it serves is not similar. The trade shop employer of a dot etcher faces an entirely different set of market problems from those faced by the integrated lithographer who employs a dot etcher of identical abilities.

In commercial lithography centered in small shops, the majority of which are combination plants, there is only one job cluster. It is made up of lithographic pressmen and preparatory crafts and may often, because of similarity of skills and identity of product markets served, include as key classifications letterpress as well as offset pressmen. Typically the skill requirements for lithographers in these plants are lower than those in color or publications houses.

In captive lithographic units such as those found in box plants, can plants or among toy makers, the job clusters are similar to those found in lithographic color houses, but in general made up of less-skilled employees, although the product markets served are identical with some of the markets served by the color lithographers. In captive lithographic units of the type known as private plants, those which supply a portion or occasionally all of the internal printing requirements of a firm, the job clusters are similar in skill characteristics to those found in combination plants, although in the smallest the one or two persons involved are swallowed up in the clerical group.

This brief summary suggests some of the strategic problems facing a union of limited resources whose intent it is to "organize lithographers wherever they may be." It also shows that firms selling in a wide variety

Table 5.   Location of job clusters.

Trade shops	Captive metal and paper firms	Color, publications and large commercial lithographic firms	Small commercial lithographers	Private plants
A—	AA	AA	—	—
—	BB	BB	—B	—
—	—	—C	CC	C—
—	—	—	DD	D—
E	E	E	E	—

Key:

High-skill cluster

A = Preparatory and platemaking job cluster
B = Multicolor press job cluster

Low-skill cluster

C = Small one-color and simple preparatory job cluster
D = Offset duplicators job cluster

Nonlithographic cluster

E = Other lithographic workforce
First letter = same skills
Second letter = same product market

of product markets may employ craftsmen in several job clusters which are closely related when found in the same firm, but which may have different wage-making characteristics when occurring separately in other types of firms. Table 5 highlights the distribution of job clusters among users of the lithographic process. It is convenient for most purposes to compress the five subgroups of workers into three job clusters for which a convenient shorthand will be the high-skill (A and B), the low-skill (C and D), and the nonlithographic (E) cluster. Although the primary feature which differentiates the three clusters is skill requirements (technology), they are also differentiated by the type of firm which employs them, and, as we shall note shortly, by the way they are organized in unions.

Suppliers of Labor

The different kinds of labor described in the preceding section are drawn from a number of sources. Historically there was only one important source of skilled labor, apprentice training within the lithographic firms. On occasion this has been dominated by the lithographic union; at other times it has been controlled by the employers. Apprenticeship is not primarily a union device to restrict entry. Its use in nonunion firms and during periods when the "open shop" prevailed attests to its effectiveness as a method of imparting skills. The lithographic industry is not unique in this respect; Sumner Slichter found this to be the most common situation in his study in 1941 of union policies.[3]

Apprenticeship is not necessary in the low-skill job cluster, for in these jobs an inexperienced worker may attain proficiency within a much shorter period than the conventional four-year apprenticeship. For such workers alternative means of training are more efficient, but are likely to be used only where the employing firm is not organized by one of the craft unions in the industry. In fairness to the unions, however, it must be pointed out that their enforcement of restrictions on the productive use of apprentices may be very lax in such cases, so that the disadvantage to the employer is slight. Among the employers interviewed during 1957–1958, none viewed restrictions enforced by the ALA by means of the apprentice clauses as a serious hardship.

Another source of trained workmen is the trade school. Trade schools are of two varieties, those which provide boys with skills, and those which provide the industry with the proper number of properly skilled employees. The first sort typically is a part of the public school system and has no connection with the union or employer associations. Although the industry does receive some workmen from this source, the lack of employer or union approval means that entry to most firms through other than the traditional feeder route is barred to the trade-school graduate, and the skills learned are therefore of little competitive advantage to their possessor.

The more important type of trade school is that which has close relations with employer associations or the union, or both, and integrates its training program into the apprentice program of the industry. Such schools operate in a number of large cities, the best-known being in New York, where the school receives support from the city, and in Chicago, where the school is jointly owned and operated by the Chicago Lithographers Association and Local 4 of the ALA. Such schools provide retaining of technologically displaced journeymen as well as supplemental training for the apprentice, and union-management cooperation in the use of such facilities has been an important factor in permitting a rapid introduction of new techniques as they have become available.

Among the other unions that organize lithographers, the only ones that accept a continuing responsibility to provide members with jobs are the printing pressmen and assistants (IPP&AU), the photo-engravers, and the typographical union. Of these three the IPP&AU is the most important. It has attempted to organize lithographic pressmen since 1913, without notable success before 1945.[4] With the growth of the combination plant, the IPP&AU, by continuing to represent all pressmen and assistants (feeders) in such firms, took in numerous offset pressmen. When such plants put in their own platemaking equipment, IPP&AU in turn organized the workmen operating this equipment.

Throughout the early 1950's it is probable that the number of lithographers in the IPP&AU grew more rapidly than it did in the ALA. The absence of IPP&AU data makes it difficult to document this, but bits of information from several independent sources show that this union is now a significant factor. For example, in the Boston area the pressmen represent offset workers in the Cuneo Press and Rapid Service Press, the former a large and the latter a medium-sized combination shop, plus a number of smaller firms. In Lowell, Massachusetts they control the large Courier-Citizen Company and a number of the medium-sized firms in that city. As late as 1959 they probably had a greater membership among offset workers in that city than the ALA. Among the larger cities nationwide, they have a substantial offset membership in Atlanta, Dayton, Indianapolis, and Washington, D.C.[5] Perhaps the most useful item is an offset press manning survey conducted by the Union Employers Section of the Printing Industry of America in 1956.[6] Of 212 plants reporting, 100 were organized by the ALA, 101 by the IPP&AU, and in eleven the offset department was unorganized. These figures are in no sense representative of the industry, for it was the unionized employers of a predominantly letterpress association that were answering the survey. Seven eighths of the reporting plants were combination houses; this, and the fact that the Union Employers Section of PIA conducted the survey, weighs the sample in favor of the IPP&AU, since it may be assumed that substantially all the plants having letterpress departments were organized by that union, or else the employers would not have been members of the Union Employers Section.

It is clear that in 1956 the IPP&AU was stronger in its offset membership than it was in 1946, though not yet a major organization in the industry.[7] Unfortunately, there is no reliable evidence of current IPP&AU strength to check the plausible hypothesis that it has continued to expand its offset membership more rapidly than the lithographers' union.

The size of the IPP&AU offset membership does not indicate its bargaining strength, for these members are concentrated in the low-skill job cluster. At present the IPP&AU probably exceeds the ALA in members operating offset duplicators, the general class of presses smaller than 17″ x 22″, and may represent more pressmen on standard size one-color presses. Without doubt it has a growing membership in the high-skill cluster, but the fact that it still advertises its offset trade school in Pressman's Home, Tennessee, as capable of turning out lithographic workmen in six weeks suggests that the typical offset worker represented by it is not highly skilled.

The photo-engravers' union has never actively attempted to organize lithographic craftsmen in the preparatory crafts. When photo-engraving trade shops have added offset platemaking in order to service customers who have put in offset presses, photo-engravers will learn the new skills necessary to produce lithographic plates. But this union has made few efforts to organize trade shops engaged solely in lithographic platemaking and has not actively assisted its trade-shop employers to compete in the lithographic trade platemaking market.

The ITU concerns itself primarily with one lithographic skill classification, that of the stripper. As photo-typesetting becomes more important as a competitor to hot-type composition, some work traditionally done by ITU members is eliminated and in part replaced by the task of pasting the film into place. The ITU is intent on seeing that the job opportunities for its members shall be preserved by continuing to include within its jurisdiction the makeready of the type preparatory to platemaking, regardless of the process used to print the copy. This brings it into conflict with both the IPP&AU, which represents some strippers because of their presence in the combination plants, and with the ALA. Like the IPP&AU, the ITU has set up trade schools for retraining its members in offset. The primary skill which the ITU seeks to teach is stripping (called paste make-up in their terminology), whereas the IPP&AU school at Pressman's Home teaches the full range of lithographic skills.

There are relatively few large firms in the lithographic industry which are not organized, though this is not true of smaller firms. In Boston, for example, the ALA organizes all five of the largest producers of lithography, but has only twenty shops under contract among perhaps eighty users of the lithographic process.[8] Another half-dozen of these shops are under contract with the allied unions, but the balance are nonunion. Few of these nonunion shops compete in the market for color or publications lithography, but they are an important factor in the market for commercial lithography. Workmen in these shops could not hold a job in one of the large lithographic plants in Boston, but they would be qualified in most plants which specialize in supplying the local market for commercial lithography. So also would some of the operators of small presses and simplified platemaking equipment found in private plants. There is of course no clear break in the spectrum of lithographic skills, but the undoubtedly large number of unorganized lithographers is to be found primarily in the low-skill segment of the skills spectrum.

Other "suppliers" of lithographic skills are the press manufacturers, who provide brief training courses along with the press in order to get it in operation. When the order is large, this training program may be extensive and supplemented by the employer with special instructions and the use of such material as that provided by the Lithographic Technical Foundation. Such, for example, was the method adopted by the Sutherland Paper Company when installing a $1,000,000 lithographic operation in the space of three years. This method of imparting labor skills is growing rapidly in importance because the lithographic process is being introduced into plants in locations where there is no supply of trained labor, and where the slight advantage of having the ALA urge craftsmen to move to the plant is far outweighed by the high cost of accepting the working conditions desired by the unions.[9] The actual number of such firms is currently not large, and in most major labor markets the ALA is still the acknowledged source of supply for workmen in the high-skill job cluster, and to a lesser but still significant degree in the low-skill cluster.

Labor Mobility

Geographic mobility has long been a quality of journeymen in all the printing trades, and there are still procedures in the current ALA constitution to facilitate the geographic movement of craftsmen:

Should there be no vacancies in [the local president's] jurisdiction, he shall forward immediately to the International President the names of those applicants who are willing to work in another jurisdiction, with their qualifications and wages demanded . . .

All vacancies in a local which cannot be filled by a Local president shall be reported at once to the Vice-President of that Region who shall exert every effort to fill the vacancy. If necessary, he shall send out a notice to the various Locals for members seeking positions outside their own Local, to communicate with the Vice-President of that Region who shall be the Employment Manager for his respective Region.[10]

The true "journeyman" is a thing of the past; few lithographers today have had any experience outside their own city, and currently the only important movement from city to city is during a strike when nonunion lithographers are imported, or as the strike extends from weeks into months and striking workers accept positions in other cities. Not since the extended unemployment during the 1930's have the once regular notices of local presidents been found in the pages of the union's journal, asking members not to enter their jurisdiction without first checking in with the local president, or warning them to stay away because of strike activity.

This has implications both for the stickiness of interregional wage differentials and for the ability of the ALA to place its members on the large amount of offset equipment installed in the last decade. For in this case, interest is in the relationship between the geographic distribution of total union membership, and the geographic distribution of increases in employment. To take a single example, the south central states had less than 2 per cent of the union's total membership in 1947, yet these states accounted for nearly 12 per cent of the industry's growth between 1947 and 1958.[11] The ALA has not been able to shift its members into these job opportunities, and the union's degree of organization in these states has fallen rapidly. Thus in areas where the lithographic process is being introduced most rapidly, the union has little to offer employers except increased costs, for it has no supply of skilled labor at its command. Because this lack of geographic mobility also is true of the membership of other printing unions which organize lithographers and is accentuated by the fact that their lithographic membership is concentrated in the low-paid classifications, lack of mobility is not a factor which significantly changes the relative advantage between unions.

Workers in the key classifications are properly called craftsmen largely because of the variety of products they must be qualified to make. Skill mobility between product markets is far more complete than the ability of firms to enter new product markets. This is not uniformly true, for some product specialties lend themselves to a different type of task breakdown within the skill classifications than do others, and it is to be expected, for example, that a dot etcher in a label house will require additional training before he becomes proficient on prestige calendar work. Nevertheless, the transferability of skills between product markets is a distinguishing attribute of the lithographic skill classifications.

Process mobility, the ease with which a craftsman trained in one process can switch to another, has been a matter of dispute for forty years. The only relevant movement is from letterpress to lithography, and this was true even during the 1920's when lithography was not growing at a faster rate than letterpress. The IPP&AU has long claimed the offset press was only insignificantly different from letterpress equipment, but other than among the leadership of this union it is conceded that not only the learning of new skills but the unlearning of old ones is required to convert a skilled letter-pressman into a skilled lithographer. Nevertheless, this lack of process mobility must be qualified in two ways. Some of the recent offset equipment requires little skill in its operation; a letterpress craftsman or another worker can be taught to operate it in a few weeks time, as has already been noted. There is in addition no reason why a person cannot learn to operate both types of equipment, given the opportunity and sufficient time. As the number of small combination shops has grown, so has the opportunity for IPP&AU pressmen to develop the craft skills required for efficient lithographic press operation. Once a number of such workmen is available in each local labor market, the apprentice system becomes an effective method of developing additional journeymen. Equally important is the fact that several of the separate operations of each printing process have in the past three decades become almost identical. This is particularly true of the preparatory stage, and it seems likely to become even more pronounced in the future.

We have already noted that the journeyman lithographer seldom travels between cities. The pattern is less uniform with respect to movement between shops in the same city. Employers tend naturally to resist movement of craftsmen in and out of the shop, because the known is being replaced by the unknown, because there is an inevitable temporary loss of efficiency, and because no loyalty to the firm is possible under such circumstances. By far the most important reason, however, is that wages are perfectly flexible only at the time of hiring, and a city with a mobile craft group is nearly always a city with relatively high wages. Some employers try to reduce the mobility of their workforce by offering delayed benefits such as pension plans and life insurance policies, end-of-year bonuses or profit-sharing arrangements, although it may be noted in passing that few have urged seniority provisions to secure this end. Historically, a measure of the strength of local or national employer associations has been their ability to secure acceptance of a no-raiding rule by their membership, the basic provision being that no craftsman may be hired at a rate higher than the rate paid by his previous employer. This was a provision rigidly enforced in many cities back in the days when the LNA was strong, and was apparently still applied in New York in the mid-1930's and in Baltimore in the late 1940's.[12] It is still on the rule-book of the Canadian Lithographers Association, according to an officer of the Toronto local.[13] The same effect is achieved in cities where employers agree to pay no more than the contract scale for all classifications.

The efforts of the local officers to secure premium payments above scales for their members vary greatly. Aggressive officers work actively to move the men around and push premium levels as far above the scale as possible. Toronto is such a local. In others the local officers are faced with one large and several smaller plants, with the membership in the large plant about equally committed to the union and the company and having no desire to move for slight increases in wages. Boston and Rochester are examples of this rather common situation. In other cities there is little opportunity to shift because the local president believes his responsibilities as employment manager for the jurisdiction compel action only when men are out of work. In such situations even trade shops may pay little above the scale rates.

Just as company-sponsored pension plans tend to reduce mobility, pension plans and other delayed benefits bargained for on a group contract basis tend to equalize nonwage payments between firms, and permit the employee to suffer no loss when he moves from firm to firm within the group. In New York, for example, even vacation credits are "vested" after five weeks of work for the same employer, and full transferability is insured by the provision that an employee who has not worked a full year for one employer is nevertheless entitled to time off for vacation, the amount paid being proportional to the amount of time worked before the vacation period.[14]

Pensions, health and welfare plans — such things are all contracted for on a group basis between Local 1 and the Metropolitan Lithographers Association, and minimize any losses to an employee resulting from a shift in employers.

Without question high mobility favors the union. It improves the union's ability to apply pressure to specific employers, and permits it to move wage levels up during the contract period as well as at negotiation time. But mobility also has advantages for the employers. A craftsman who has worked in a variety of plants has had a variety of experience, and it has already been noted that one of the most highly prized attributes of the craftsman is his ability to work efficiently on many different types of products. Moreover, high labor mobility permits an efficient distribution of the labor force, even when rates are relatively inflexible downward. The question of which party, union or employer, benefits most from mobility within a city depends on how the individual premium payments are handled in contract negotiations. A $5 a week increase to everyone leaves individual premiums untouched, whereas a $5 increase in scale rates means that workers receiving premiums will have that much of their premium "sopped up" in the wage settlement. The premium rates do not move down, but they do remain static while the floor, the scale rates, move up. Consideration of this question is taken up in a later chapter.

One final type of mobility should be mentioned, the ease with which craftsmen may move from the ranks of employees to employers. The owner operators of four of the twenty largest lithographic firms in Boston began as craftsmen, and there are very few owners of lithographic trade shops who began any other way. In some localities these men are allowed to remain in the union, and on occasion may be active in local union affairs. At least three locals have had presidents who were employers, not employees. In Chicago, three trade shop employers who are members of the union are also on the employer bargaining committee. Craftsmen may become employers; they may also in a large shop be raised to a supervisory level. The ALA has no fixed policy toward such members. In Rochester they are normally denied the right to continue membership; in Toronto they are in trouble if they do not continue; and in Boston the local has no strong feelings either way. In the past, such issues have been a minor source of difficulty for the union but have never led to fixed positions such as are found in the ITU and the IPP&AU.

Labor mobility and wage levels. It would be convenient if one could illustrate graphically the way in which the different degrees of mobility, levels of skill and types of market pressure affect the results of collective bargaining. It is not possible to do so, and attempts to plot supply and demand curves for labor against some sort of labor price and quantity axes lead to muddled thinking in problems of lithographic wage determination. The reason is that the two elements of central interests, the price for and the amount of labor sold in the market, are not simultaneously determined.

There are essentially two reasons for the independence of these supposedly interdependent results. (1) Wage determination is primarily a short-term process, while the factors most important in determining the amount of labor required operate over a longer time span. This may be said in a different way. No one can specify the upper limit for a wage increase that will not materially affect employment opportunities, for no one knows how much of it can be absorbed by decreased profits, increased managerial efficiency, docile customers, increased worker efficiency, or neutralized by encouraging workers in competitive plants to secure like increases, or by the continuing changes in the production process or in market opportunities. Most important, no one knows the time span over which each of these counteracting (or accentuating) forces may become effective. (2) Geographic mobility of workers, particularly in the low-skill and nonlithographic clusters, is very limited. But if workers do not shift from area to area, wage pressures do. Of great importance are wage pressures which arise in the product market. So long as it remains easier to ship goods than to ship workers from place to place, one must expect wage pressures and equilibrating adjustments in the supply of labor to be out of balance. Needless to say, the fact that some lithographic product markets are local, some regional, and others national, would complicate any graphic analysis materially. Moreover, not all wage pressures come from the product market. What has been called the "orbit of coercive comparison" can be coercive merely because of the way workers are organized into unions. It is an oft-repeated dream of the ALA that wages, hours and working conditions will one day be identical in all major locals. In a strong union, such dreams affect wage decisions.

Chapter **VI** ✆

The System before the Thirties

Early History of Union-Management Relations[1]

The earliest records of employer and union organizations in the lithographic industry appear in 1853. Hoagland and Foner[2] refer to an organization of lithographic workmen in New York City, and in 1856 there are references to a Lithographers' National Union. Perhaps as a result of a strike of lithographic printers in New York City in 1853, there was formed in that year the New York Lithographic Employers' Association.[3] The constitution of this body required all members to set a ten-hour day for their workmen, employ no runaway apprentices, and be honest when asked for information about former employees or customers. Nothing is known of its membership, nor how long it existed. There is little additional information of either union or employer organizations until 1870, when there is again a report of a local employee organization in New York City, which according to Hoagland "became the nucleus of the first national union in the industry, formed under the Knights of Labor in 1882." [4] Connection between the organization formed in 1870 and the one in 1882 is very tenuous, if it exists at all. Union sources show that eighteen lithographers employed in a Jersery City plant formed the "Romar Fishing Club" in April 1882, almost immediately changed the name to the Hudson Lithographic Association, which on June 10 of the same year was chartered by the Knights of Labor.[5] It is this union which the present-day Amalgamated Lithographers of America claims as its oldest direct predecessor, and 1882 as the year of its birth.

The Hudson Lithographic Association was made up of transferrers and pressmen, with none of the creative crafts such as artists or engravers represented. Its concerns were the practice of having a pressman run two presses, "teamwork," or having a transferrer use semiskilled helpers in limited parts of the job, and having employees work overtime at the straight hourly rate. As the union expanded in the New York City area, its central aim shifted to reducing the hours from ten to nine per day, the prevailing hours for the artists and engravers. In 1886, while still in the Knights of Labor, the union struck to enforce the shorter hours. Timed to coincide with the eight-hour struggle of the Federation of Organized Trades and Labor Unions, the strike is known only as having started on May 1, 1886, and ending soon after with no success. Of so little note is this first recorded strike of the lithographers that three years later no hint that it ever occurred appears in the first annual convention proceedings of the employers' association.[6]

As a result of the strike, the Hudson Association left the Knights of Labor and in the following year changed its name to the Lithographers' International Protective and Insurance Association, broadening its jurisdiction to include artists and engravers, and also the lesser-skilled stone grinders.[7] Few of the artists and engravers were seriously interested in joining an association which was still trying for the nine-hour day, and in 1890 these crafts established their own organization, the International Lithographic Artists and Engravers Insurance and Protective Association of the United States and Canada. In 1898 the stone grinders formed their organization, the Stone and Plate Preparers Association, and, like the artists and engravers, took with them some members formerly in the senior union. Since 1894 the original organization had been called the Lithographers' International Protective and Beneficial Association of the United States and Canada (LIP&BA), the name it continued to carry until amalgamation in 1915. In 1899 the poster artists left the Artists and Engravers Association to form their own Poster Artists Association of America, and a year later several local feeder organizations banded together to add a further impressive title to the list of national lithographic unions, the International Protective Association of Lithographic Feeders of the United States and Canada. To complete the roster other unions should be mentioned, among them the paper cutters, the lithographic embossers, and the music engravers, but most of them were restricted to a single craft in the New York City area, and none played an important role in the labor-management relations of the industry.[8]

In the 1880's employers in the lithographic industry made two abortive efforts to establish a national association. They were not successful until 1888, when they formed the National Lithographers' Association.[9] The important questions did not involve the union; the securing of a substantial tariff increase was of greater interest, and still more pressing was the allegedly growing practice of cutting prices until they barely covered wages:

The reward justly due to extraordinary labor, skill, and capital invested is slipping gradually from our grasp, through methods and practices which must prove utterly ruinous if persevered in.

A natural ambition at enlargement of their establishment has led many, in the times of prosperity, to invest their earnings in additional machinery and materials . . . [Later], rather than having any part of their operating force remain idle, work was taken at any price, and a wild and insane competition was thus engendered.[10]

Often, said the NLA's first president, Julius Bien, workmen went into business for themselves, getting machinery and material on time payments and thus being forced to work at any price, just to meet the interest and installments. And, while prices were going down, the practice of offering "extraordinary and unwarranted remuneration" to steal the best workmen from other employers was pushing costs up. In this dismal state of affairs President Bien could still see one bright spot:

Nothing has occurred to disturb the happy relations existing between ourselves and our workmen for many years. As a class of more than the average in intelligence, industrious and faithful, taking a pride in their work, [any reasonable request of theirs] . . . should receive proper attention.

The convention of the Lithographers' International Protective and Assurance Association of the United States and Canada, held at Buffalo, March 29, 1889, unanimously expressed the wish that the working hours in the printing, transferring and proving departments, should, from the first of January next, be uniform throughout this country and Canada, and to be fifty-three hours per week . . .

This request should . . . be promptly and cheerfully granted.

The request was not "promptly and cheerfully granted" by a large number of employers, for the NLA had not yet devised any means of requiring compliance from its members. Hoagland points out that not until 1906 were such procedures introduced. During the 1890's the national association "merely held its annual conventions and discussed matters of interest to the trade." [11]

It is noteworthy that during this early period of weakness in employers' organization the union itself exerted no great pressure on employers. In 1895, six years after President Bien of the National Lithographers Association made his address to the employers' convention, the president of the union touched on the same subject of shorter hours:

In justice to employers who have granted the 53 hours, we should urge upon those few who still persist in working their shops longer, the injustice of this action. As the Employers Association, of which most of those who are violating this rule are members, has frequently passed resolutions favorable to the 53 hours, I believe there should be no difficulty in gaining this point.[12]

This era of good feeling was interrupted when disagreements between the artists and their employers in New York City in 1896 caused the

Artists', Engravers', and Designers' League to place a series of demands on employers throughout the country, the most vital being the abolition of piecework and the introduction of the forty-four-hour week.[13] Moreover, the dominant poster house, a recent combination of nine separate firms, had in 1894 introduced piecework to the art department, and the artists' and engravers' union was anxious to arrest its spread to the rest of the industry. A settlement of the dispute favorable to the union was secured in all but the key cities of Buffalo and New York, where the employers had banded together into an effective organization headed by the same firms that were dominant in the national employers' association. A five weeks' strike was called, and resisted with a vigor which surprised both parties. It ended with a standstill agreement providing for arbitration of the disputed issues. The arbitration award, given by Bishop Henry C. Potter of New York, favored the union in all but the workweek hours, which were left at forty-seven and one-half, the standard for artists and engravers in the association houses at that time.

By 1900, hours, except for the lower hours of artists and engravers, had been standardized at fifty-three, usually on the basis of a nine-hour day. Wages varied a good deal between locals, but Table 6, drawn from a report submitted to the LIP&BA convention in 1901, gives an indication of the general level of wages at this time.[14]

Table 6.   Minimum wage set in 1901 convention.

Category	Average weekly wage	Minimum wage
Artists	$30	—
Engravers	19	—
Transferrers	21	$20
Pressmen	21	20 to 35 varying with size of press
Provers	22	20
Stone Grinders	12	—
Feeders	9	—

The problems of two-press operation and teamwork by transferrers had largely been solved, and time-and-a-half for overtime had become standard in all but four of the twenty subordinate associations.[15] In general, the LIP&BA membership could agree with General President Keogh that "as an organization, we have everything to congratulate ourselves on." [16]

About 1896 the loosely knit National Lithographers' Association split into two separate organizations.[17] The midwestern group, Lithographers'

Association (West), gave most attention to prices and costs and dealt with labor questions only incidentally, while in the other, the Lithographers' Association (East and Pacific), the opposite emphasis was given. Stronger unionism in the east, where two thirds of the LIP&BA membership was located and the demonstrated failure of previous attempts to regulate prices and costs were responsible for this, as was the fact that the eastern firms were assertedly more efficient than their western competitors.

The New York employers who had organized to resist the artists' demands became the nucleus of the eastern group, which soon began to develop a more active resistance to what it considered to be the unreasonable demands of the unions, in particular to those of the LIP&BA.[18] It was especially interested in regaining some measure of control over the supply of labor, which had increasingly come to rest with the five unions and their loosely knit Central Lithographic Trades Council. An unsuccessful employer effort was made to assert control over the making of apprentices, and a rule was passed prohibiting a member from hiring a craftsman employed by another association member at a wage higher than that which he had formerly received. The rule was aimed at the raiding practices of employers, but more directly at the union practice of shuttling members around from employer to employer and city to city, each time at an increase in rates. In criticizing the S. A. 1 (Subordinate Association 1, New York City) for setting a local minimum higher than that provided in the constitution, President Pritchard pointed out that this, in conjunction with the employer association rule, excluded other members of the association from working in New York City.[19] The rule involved here had some effect, but application by a limited employer group was potentially more harmful than beneficial and added to the desire for cooperation between the two national associations in the field of labor relations. This was effected in 1902, largely through the efforts of A. Beverly Smith, who was appointed in that year as the first full-time manager of the eastern employers' association and dominated its labor policy for the next four years. Beverly Smith proposed to implement the employers' desire to have a greater share in the rule-making by a system of "mutual government," which he later described to a union committee at one of its conventions:

> Mutual government consists of two parts, joint action and arbitration . . . Those things that have heretofore been settled under old conditions, by either the employers or the men, meeting alone, enunciating an ultimate resolution and then fighting to force conditions . . . those things we place in the hands of equal representation from the employers and employees.

Your constitution and by-laws — if you realize it you have no right to make any law — you can put what you please in black marks on white paper — so can we — but you have absolutely no moral right or trade right to put a line in there that governs the employer. Your jurisdiction ends at that door [indicating door of Convention Hall]. Our jurisdiction ends at our door. When we get out into the passage-way, which is the trade, we will settle our differences there jointly . . . You bring that book [indicating the constitution and by-laws] out there and say that governs the passage-way of the trade, and we say "no," no more than the constitution and by-laws of the church govern the trade. When you attempt to enforce it there, that is when we need mutual government . . . We give you half the right and demand the other half. That is mutual government.[20]

The 1904 National Agreement. "Mutual government" was not acceptable to the five unions, working now in moderate harmony in the Central Lithographic Trades Council. Their own internal frictions were matched in the employers' group where the recurrent disagreements between the New York-based and Chicago-based groups hindered effective cooperation. But the opposition of the unions to sharing decisions over working rules, centered in the dominant LIP&BA, was sufficiently irksome to the employers to bring them together, and by late 1903 they were acting in unison throughout the country to force the LIP&BA to accept a contract with a no-strike, no-lockout clause, and binding arbitration. The employers met the delaying tactics of the joint committee of the unions with an ultimatum in early 1904 that unless an agreement was reached on or before March 15, "establishing arbitration and eliminating the possibility of strikes and lockouts . . . the employers will proceed to deal with their employees individually and not through the union." [21] Active negotiation between the Central Lithographic Trades Council and the combined Lithographers' Associations (East, West and Pacific) resulted in a narrowing of differences but no solution, and on March 15 the employers mailed each of their employees an individual contract, the best ones receiving in addition a promise of five years' guaranteed employment. The question of union membership was not central to the issue, but the question of who would determine working conditions was. Employees who refused contracts were locked out by their employers.

None of the unions was financially prepared for the lockout, for it was the first national work stoppage since the industry had become well organized. Nevertheless, the membership of all the unions held firm, including the feeders, although President Coakley of the feeders' union later pointed out with some asperity that one of the concessions demanded by the employers and refused by the LIP&BA, the relaxing of apprentice provisions, would have been of direct benefit to the feeders.[22]

At the end of four weeks in which the industry produced almost nothing, both sides were ready to call a halt. The compromise agreement included as its central feature a provision for joint determination of some issues, with a provision for binding arbitration, but did not introduce "mutual government" for all terms and conditions. Nevertheless, the unions clearly lost ground in the encounter, for they formally conceded that they no longer had the right to determine working conditions unilaterally.

The Great Strike of 1906. The LIP&BA at this time was experimenting with its first non-New York president, John W. Hamilton of St. Louis. President Hamilton was proud of both the 1904 agreement with the employers and the working arrangements with the other lithographic unions — sentiments not shared by a dominant faction in the union led by the New York local. The 1904 LIP&BA convention resolved to demand an eight-hour day in the next contract, but when the other lithographic unions refused to support the demand, President Hamilton and the general officers decided not to press it to a strike.[23] This further alienated the New York wing, and in 1906 a special convention of the union was called for the purpose of writing an agreement, specifically denying to any committee or group of officers the authority to negotiate a settlement. Earlier requests of the employers' association to negotiate an extension were formally ignored by the union and instead the combined Lithographers' Associations (East West and Pacific) were summoned to the convention. Neither chose to attend. After further postconvention maneuvering within the LIP&BA and between the lithographic unions had failed to produce a common union position, the LIP&BA announced its intention to go back to the old system of working without agreements.

This prospect was sufficiently distasteful to the employers to bring them together in the National Association of Employing Lithographers (NAEL). Each member agreed to place all questions concerning the total conduct of his firm's employment relations in the hands of the association, and to take no action in this field without its approval. The ultimate enforcement mechanism was a high initiation fee, $500 per press, for which the member signed an undated note. All things considered, the new association was closer in structure to an international union than to the employer groupings it replaced.[24]

The organizing meeting of the NAEL was held in May 1906, and in early June the new constitution was adopted. Barely a month later the

LIP&BA independently announced to the employers that the forty-eight-hour week would go into effect on September 1, and if the demand was not conceded the union would strike on August 1. This demand destroyed what little remaining unity there was in the Central Lithographic Trades Council, and for all practical purposes the organization ceased to exist.

The NAEL then suggested negotiations of the hours question between "committees of the two organizations with full power." [25] The employers probably knew that negotiations under these rules were politically impossible for the LIP&BA, whose leadership could not commit the union to a contract without a referendum vote by the membership. With the breakdown of negotiations, the NAEL put into operation a well-planned and coordinated open-shop campaign against the LIP&BA. The other unions tried to maintain neutrality, and quite possibly could have done so if the employers had been anxious to keep them out of the fight. But this was not the case. Said the NAEL:

The Board of Directors now feel that the lithographic trade is about to "shake off its shackles" and to emerge from the present struggle with conditions that ought to prevail at all times, and in which the man who owns the shop is the man who controls it.

Eventually, all but the poster artists became involved in the strike, affecting, at a conservative estimate, some twenty-five hundred employees.[26] As each of the other unions was drawn into the fray, the NAEL declared "open shop" against them also. The individual contracts perfected in the 1904 lockout were again used, this time without the saving provision that they would become void if a settlement was reached with the unions. Some were straight yellow-dog contracts, like the example which began:

You represent to us that you are not a union man and agree not to hereafter join any union without our written consent. In consideration of this representation and the performance of the promise, we hereby employ you as a _____ for a period of _____ years.[27]

Life insurance policies, bonuses payable at the end of ten weeks of uninterrupted attendance, or bonuses for bringing in new men were among the inducements employed, and were buttressed by the requirement of written resignations from the union, the sending of loyal employees into union meetings "undercover" to help break the morale of the strikers, and the use of information gathered by private detective agencies.

As the strike extended from weeks into months, both individual employers and groups of union members became less intransigent in their position. But the NAEL leadership, with full power to conduct the strike, was determined that it not be permitted to end in compromise, or indeed in anything short of the effective elimination of the union:

> It would be a calamity to the lithographic trade if the strike were to break at the present time and the men were permitted to come back to work under the open shop system [hiring both union and nonunion men]. If the strike were to end today there might be a repetition of the present strike next year or the year after next. We want the strike to last long enough to enable us to sign contracts with 35% of the men, and then the open shop is here and can be maintained; and another strike will be an impossibility. Make individual contracts mean a preservation of the open shop.[28]

On the union side the international leadership under Hamilton was eager to find a settlement to a strike it had feared since the start of the eight-hour agitation, and in New York it found two large firms that were members of the NAEL, Sackett & Wilhelms and Werther-Rausch, who were equally anxious to resume production. An agreement on the gradual reduction of hours was signed between the international and the two firms, but repudiated by the New York local. But it was not rejected by all the workmen concerned, and President Hamilton "toured the country" with the vice-president of Sackett & Wilhelms, rounding up eighteen additional employees who would be used to staff the plants which S. A. 1 was still striking.[29] S. A. 1 initiated impeachment proceedings against the general officers; and the international responded by setting up a new local, S. A. 28 in Brooklyn, specifically for the two firms, and finally expelled S. A. 1. The two firms were in turn fined by the NAEL, their "notes payable" presented for immediate payment, and then hauled into court. The outcome of the legal battle was not determined until late 1907, well after the conclusion of the strike; the NAEL lost on the first round, and then appealed and collected on the second.

Quite possibly as much as two thirds of the members of the five lithographic unions were employed in association shops, and therefore locked out.[30] The stiff assessments on members still working — 25 per cent of salary in the LIP&BA — soon resulted in the loss of more members than the minimal strike benefit they were permitted was worth, and the striking members were left without benefits other than those provided by the subordinate associations. In March 1907 *the National Lithographer* estimated that the struck shops were working at 50 per cent of capacity, and in May the LIP&BA, the only one of the unions other than the poster artists that still had a working national organization, sent out for referendum the question, "Shall the strike be called off?" The outcome of the vote was a foregone conclusion, and in June the LIP&BA, now concerned only with holding the remnants of the organization together, informed the independent shops that they could go back to fifty-three hours.

Little is known of how the other unions fared. They continued to exist, though probably their membership was restricted to independent shops. The NAEL, with every shop an open shop, was troubled mainly by members who thought the ghost of unionism had been permanently laid. The long strike had made all the employer members more or less equally antiunion, and the devisiveness which had plagued early efforts at unity was greatly lessened. On January 1, 1911, NAEL instituted the eight-hour day on its own initiative, and in the face of resistance from some member firms. But in the words of the NAEL president:

It was in part a question of good judgment as whether or not it was worth more to have the eight hours and no union, or to continue to have nine hours with the union growing up again and overthrowing the open shop and demanding the closed shop.[31]

By retaining full control of the labor relations function in all member firms, NAEL was able to keep track of employees partial to unions, insure that all employers maintained their full quota of apprentices, and insist that employees be treated with sufficient fairness so that there would be no eruption of unionism in the industry again.

Not all lithographic employers had been willing to hand over so much of their freedom of action to NAEL. A number of large employers had historically been outside the association. In 1904, for example, not a single Philadelphia employer was in either the eastern or western group.[32] At the inception of the strike NAEL claimed that it had 75 per cent of the industry's productive capacity in the association, and that another 10 per cent was in firms cooperating with it. These are overstatements, but to what extent, it is difficult to say. By 1915 NAEL probably had declined somewhat in relative strength, but it still controlled at least 50 per cent of industrial capacity.[33] Yet there is no question that from 1906 to 1919 NAEL was the decisive voice in determining wages and working conditions in the lithographic industry.

Birth of the Amalgamated Lithographers of America

In spite of the repeated failures of the five lithographic unions to find some basis for continuing cooperation, there continued to be efforts toward this goal. For it was a goal, both for a significant proportion of the membership (a proportion which increased substantially after the great strike), and for some members of the leadership in each of the unions, most notably the lithographic engravers and designers. In 1912 a conference was held among the locals in Chicago in which "all crafts were represented";[34] it is not clear whether all unions were. The conference was the first which did not end in total disagreement, and it led to another in 1913 at Peterson's Hall in Buffalo.[35] In four days of sometimes acrimonious discussion, representatives of the five unions worked out a proposal for amalgamation that was acceptable to all, but pleasing to none. The four smaller unions accepted complete amalgamation, the LIP&BA accepted a general executive board chosen on a craft basis, but with a union-wide electorate, the unwritten understanding being that pressmen would nominate the pressman representative, the designers, the designer representative, and so forth. Herman Kaufman of the Artists', engravers', and designers' League[36] proposed that on all "basic questions the new organization would be supreme . . . but all questions relating to working time, apprentices, wages, and all other similar questions are to be under the control and jurisdiction of the membership of the craft so affected." This in turn was qualified by Frank Gehring of the LIP&BA so that it would not mean that any branch could involve the whole association in a strike over such reserved items, and again qualified not to rule out the possibility of emergency strikes. After some discussion the whole was passed unanimously. In this way problems which had kept the five unions apart since the turn of the century were gradually resolved. High initiation fees were proposed by the LIP&BA, opposed by the rest and dropped to ten dollars. A further reduction of five dollars was proposed in the first meeting of the international council, and approved in referendum. The provision for compulsory participation in death benefits raised difficulties and was finally made optional over the protests of the LIP&BA which correctly foresaw that this would create friction at the local level.

Perhaps the best evidence that the basic differences had been compromised lay in the agreement of the conferees to strike out the requirement that all votes of the international council [37] be unanimous. It was replaced, however, by a two-thirds requirement for most types of questions, not that of a simple majority. It was an extra piece of insurance against the LIP& BA's domination of the new union, and it proved fairly effective since the craft representation on the council made it highly unlikely that the LIP&BA would ever provide nine of the thirteen members. The refusal by the feeders and poster artists to join, and the addition of two vice-presidents somewhat altered the actual make-up of the council from that anticipated by the Peterson's Hall conferees.

In voting on the proposed amalgamation, the poster artists and the feeders rejected the proposition, and so the Amalgamated Lithographers of America was born on January 14, 1915, with the following membership:

3101 members from the Lithographers' International Protective and Beneficial Association (LIP&BA).

543 members from the International Union of Lithographic Workmen (IULW), formerly the Lithographic Engravers' and Designers' League.

146 members from the Stone and Plate Preparers' Association (S&PPA).

3790 Total Membership.

The feeders union, after some delay due to tactless interchanges, joined in 1918, but the poster artists remained aloof. Further discussions in 1919 ended when the ALA leadership realized that the poster artists wanted not only complete autonomy, but a veto power over any ALA actions that would affect them. Sam Maitland, the midwestern vice-president of the ALA, was barely coherent when he commented on the discussions:

> Poster Artists — the most arrogant lot of men, the biggest lot of men holding their head high — so high that we cannot get anywhere near them, as it were — a sort of an honor to sit in conference with them . . . I want to say that this resolution [against the poster artists] is not half strong enough to suit me.[38]

The poster artist association's strength was sapped the following year in an unsuccessful strike, but it nevertheless continued its independent ways, until in 1942 the remnant of a once powerful organization came into the ALA with the sole concession that the initiation fee be waived.

Organization of the ALA. The distribution of power in the ALA among the president, the council, the convention, and the membership had under the Chicago resolutions in 1912 been heavily in favor of the members. The Peterson's Hall conferees recognized the impracticability of such a power distribution, and provided that more power should rest with the convention and to an even greater degree with the international council. All but two of the positions on the council were in the nature of reserved seats, the vice-presidents of their respective regions and the nine representatives of their branches of the trade. The major distinction between the president and other council members was that the president chaired the meetings and was, like the secretary-treasurer and first vice-president, a full-time paid officer who was expected to spend most of his time organizing. This also was the major duty of the first vice-president.

It is somewhat difficult to describe the prevailing attitude toward the president since he was looked upon as both a leader and a hired man. As a leader he was respected for the trust shown in him by his election, distrusted because he might misuse power to favor his craft, his city, or to build up his own prestige and power. As a hired man he was courted because he could help build up local unions through his organizing efforts, excoriated when he could not deliver the goods.

This concept of the president's role was illustrated in the first convention of the ALA in 1917. Delegates from the Cleveland and Detroit locals were dissatisfied with the attention they were receiving from the first vice-president and moved to require national officers to come to a local whenever the local thought it necessary. Portions of the ensuing debate are given below:

National President Bock: What is your pleasure now? You have reconsidered, and now you are wrangling, you don't know what we are about. Don't you think the words "should it be deemed necessary" are sufficient?

Bro. Imhoff: By the Local.

Bock: Will the Local also pay the expense to the National Office if they call him on a fool's errand?

Bro. Simmendinger: [Denver, third vice-president — an unpaid office] I think the point that has been made is very well taken. It seems to me that the National Officers are very much afraid that they are going to spend fifty dollars railroad fare trying to settle some difficulty even if it is imaginary, they will let the thing run on till they have a strike on their hands, and they spend $14,000.00 to cure the strike . . . We are paying these National Officers to do certain work . . . We don't expect to pay an Officer a salary and then have him sit back. I think the time has come when we should insist that they work for this organization, and if they draw their salaries that they should get out and hustle just like the rest of us do in the shop . . . We want action for our money just like anybody else does.[39]

It was several years before the membership could be convinced that placing the national officers under the direct authority of thirty-odd local presidents was neither workable nor wise. As the necessity for freedom to act quickly in emergencies became evident to the ALA, without the delay and publicity attendant upon a referendum vote, authority did become more centralized. But it was given to the council, not to the president. It is central to an understanding of the union and its leadership during this period to recognize the vitality of the democratic ideal, and how strongly it affected leader-member relations.

The Offset Press Controversy

Amalgamation of the three lithographic unions in 1915 and the addition of the feeders in 1918 eliminated many of the jurisdictional questions which had troubled the industry in these early years, but not before another and more serious question arose over job rights to the offset press.

The introduction of the offset press in 1906 marked the first innovation in lithography which gave that process an opportunity to compete in product markets which until then had been exclusively in the letterpress domain. Its introduction was rapid and well advertised, and it occurred at a time when the lithographic unions were recovering from their total defeat in the 1906–1907 strike.

The printing pressmen were at this time changing presidents. George Berry was elected in 1907, and two years later began to urge his membership to gain control of the offset press. He promptly secured the revision of the IPP&AU constitution so that it included all offset printing, and initiated plans for a trade school to teach members how to operate the offset press. This was established in 1911. The LIP&BA raised a complaint before the AFL, but was no match for Berry in the political infighting of the federation. In 1915 an adjustment committee of three was chosen by the parties (a barber by the printing trades, a patternmaker by Gompers, and a fireman by the lithographers), and in a unanimous decision awarded jurisdiction over the offset press to the IPP&AU, and jurisdiction over offset platemaking to the IPEU (which union had joined the dispute two years earlier at the invitation of President Berry). In 1918 a plan of amalgamation was offered, which in essence was that pressmen and transferrers would be taken over by the IPP&AU, and the preparatory crafts by the IPEU. Having just completed a joining of the lithographic crafts into the Amalgamated Lithographers of America in 1915, the lithographers were somewhat less than enthusiastic at the prospect of destroying their organization. Moreover, they still could not understand what had happened. Throughout, they had argued that the amalgamated had all the lithographers (or offset workers, as the IPP&AU called them), and the IPP&AU and IPEU had scarcely a half-dozen. What they never understood was that *who* had them was irrelevant, at least in the case of a small union. The decisive issue was: Who *should* have them in order to avoid dual unionism? While the ALA spent its time amassing statistics to show that of all known offset presses it controlled 80 per cent and that the balance were nonunion, Berry spent his time asking the patternmaker, the barber, and the fireman if they could see any difference in offset and relief presses, challenging them to identify which picture was produced by which method of printing, and pointing out the grievous inroads which the offset press might make in the letterpress industry's market. When the lithographers refused to split themselves up between the two printing unions, the 1919 AFL convention agreed to expel them when the IPP&AU and IPEU would request it. The two unions did not request it, for the minor advantages of having a rival union expelled were already theirs by virtue of the 1915 adjustment committee award. During the 1920's the dispute took its place as a gracefully aging and insignificant contribution to the federation's list of unsolved jurisdictional conflicts.

Union-Management Relations, 1915–1921

There were no formal relations between the NAEL and the ALA for over a decade following the 1906 strike. NAEL member firms successfully defeated the union in two strikes in Cleveland and Detroit in 1915 and 1916, but country-wide the new union prospered. There was no serious problem of unemployment; on the contrary, the ALA was able to place an increasing number of members in association shops, and gradually gain back those members that had been lost in 1906–1907. The efforts of the NAEL to find workmen during 1918 prompted a general letter from the union, reminding all members that they had the opportunity and the obligation to secure jobs through the union.[40] The five-year contracts that had been given to employees during the great strike were quite generally allowed to lapse, and by 1917 the ALA was of sufficient stature in the industry to enable the NAEL's executive secretary, P. D. Oviatt, to accept an invitation to address the convention.

Why did NAEL allow its secretary to recognize the importance of the union formally by accepting an invitation to address its first convention? Scarcely ten years earlier it had been the fixed policy of the association to eliminate all unions, and to convince the workers that unions were useless. Lack of access to the historical records of the association makes it necessary to guess at the reasons, but there would seem to be two. The first quite simply was that the ALA had partial control over the supply of skilled labor. When such labor was not in demand, the incompleteness of the control made it largely ineffective as a bargaining weapon. When there was a distinct shortage, as there was from 1916–1920, this control could be used to punish firms which refused to recognize the union by refusing to supply workers, and by moving members already in such shops to others which did recognize the union.

The second reason may be equally important. Ten years of labor peace had surely allayed some of the bitterness generated in the strike of 1906–1907, and thoughtful employers could recognize that the ALA of 1917 was a totally different union, with a far less imperious attitude, from the old LIP&BA before 1906. Particularly in the large centers, it must have seemed poor policy not to deal with a reasonable union that included a majority of skilled workmen, particularly when such workmen were in short supply.

National contracts, 1919–1921. The ALA was no longer contemptuous of collective agreements. President Frank Gehring had warned the new union against them in his farewell address in 1915[41] but the rapidly rising prices of the war period and the considerable variation in the ability of locals to match these with wage increases made a uniform agreement seem very desirable to the new leadership. A letter of unusual tact was sent to the NAEL, matched in its courtesy only by similar letters going from the employers' association to the union in the 1902–1904 period. For the situation was much the same, except that now it was the union that was requesting permission to share in the rule making.[42] The letter referred to Secretary Oviatt's speech to the convention, which "found a particularly responsive chord in the breast of every delegate present," and, by instruction of the convention, the union officers were requesting negotiations:

to the end that harmony and good will together with a friendly understanding may exist between lithographic employer and employee, realizing as we do, how imperative unity of thought and action are necessary amongst all classes during the present crises through which our common country is passing.

The response of Secretary Oviatt was neutral. In the 1911 NAEL convention, Vice-President Clothier had said that "the spirit of 1906 still pervades our ranks, and we are one — absolutely." [43] But by 1917 the traditional rift between unionized employers and antiunion employers was again becoming evident, and a number of employers, notably in Chicago, where the union was taking full advantage of the tight labor market, were anxious to stabilize relations with the union. Several incidents nearly caused the breakoff of negotiations, but the advantages to both sides were clear enough to cause the NAEL to decide in its 1919 convention that an agreement would be desirable if it included the following eight points:

1. The open shop should be preserved (as opposed to the closed shop or completely nonunion shop).

2. A reasonable apprentice ratio flexible in operation should be established.
3. A no-strike and no-lockout clause, with provisions for binding arbitration should be included.
4. Reasonable guarantees of enforcement of the contract provisions should be provided.
5. Only the collective agreement would be binding; the constitution of either association would have no force in the shop.
6. Individual contracts with employees would still be permitted.
7. Any wage system that would lead to greater productivity would be permitted.
8. The 48-hour week would continue in effect.

The actual agreement concluded in August of the same year left all points except wages and hours untouched.[44] Wages were raised by five dollars across the board on October 1, 1919, over the rates paid on July 1, 1919. There were to be no further increases for one year, and a national wage commission was set up to determine the increase at that time. Wages were not a subject of arbitration.

On the question of hours, the associations had conflicting mandates — the NAEL to continue the forty-eight-hour work week, the ALA to get forty-four, on a date to be determined by negotiation with the NAEL. The compromise was embodied in the following paragraph:

Forty-four hours as a basic work week shall be adopted on May 1st 1921 or earlier, provided at that time that the basic work week shall exist nationally in the typographic trade.

The ALA membership approved the agreement with some grumbling over the small pay increase; the NAEL resolution containing its approval included a further resolve that the eight-point creed be reaffirmed, with the exception of Article 8.[45]

During the following year the cost of living rose precipitantly, the June 1920 figure being 24 per cent above the December 1918 level, and about 16 per cent above the price level during the 1919 negotiations.[46] But the agreement served to keep organized unrest at a minimum during the year, though by early 1920 the pages of the *Lithographers Journal* were sprinkled with criticism of the 1919 agreement.[47] There was much cause for criticism because since the formation of the NAEL in 1906, lithographic wage rates had fallen steadily behind those in the letterpress trades. From 1917 to mid-1919 alone, the $11.50 per week increases of the lithographers compared very poorly with the $15 to $16 per week increases which the letterpress trades, including the press assistants, received.[48] As ALA Vice-President Maitland told the 1919 convention delegates:

I don't know a trade that I can throw a lithographer out into, that has a cheaper level. I used to get out that old expression "Quit the business and carry the hod." But the hod-carrier has got you skinned to death. So I don't know where to put you now.[49]

In June 1920, a new agreement was signed, granting a flat increase of 20 per cent on current wages, unless prior increases had been collective ones, specifically anticipating a future negotiated increase. The old contract language was extended, and a provision added that grievances were to be processed nationally, not locally. Once again, the agreement was accepted by the membership of both organizations, and went into effect in July, 1920.[50]

The 1922 Strike. The lush years following World War I were ended by a sharp depression which reached the printing trades in the latter part of 1920. Early in 1921 the president of the NAEL publicly informed the secretary of the open-shop Employing Printers' Association that the lithographic industry would under no circumstances introduce the forty-four-hour week if it was not generally introduced by the other printing trades on or before May 1, 1921.[51] This, as he took pains to point out, was strictly in accord with the 1919 agreement, as extended in 1920. The public position of the ALA leadership was to condemn this double-dealing as a typical example of employer perfidy, a position apparently dictated less by conviction than by the need to explain former unqualified assurances to the membership that the forty-four-hour week had been gained at the bargaining table.[52] In spite of the fact that the other printing trades were striking for the shorter hours, the ALA made no effort to enforce the forty-four-hour week, and strikes or lockouts prior to January 1922, were local affairs revolving mainly around resistance to wage cuts (largely in non-NAEL plants) and attempts to establish open-shop hiring and shop practices. In early 1921 the NAEL suggested that wages for the next contract year be adjusted by applying the same cost-of-living argument that had been used to increase them at each of the two prior agreements. The cost of living was clearly falling:[53]

June 1920	100.0
December 1920	92.5
June 1921	83.0
December 1921	80.5

So also were sales. A "buyers' strike" was referred to in the *National*

Lithographer of February 1921, which was apparently of short duration, but it crystallized sentiment in the employer group in favor of a wage cut.

The first attempt of the wage commission created by the national agreement to decide the extent and direction of the wage change for the year 1920–1921 ended in failure. In September the NAEL in convention ordered the employers members of the wage commission to try once more to reach an agreement that would reduce wages to "reasonable levels," which, probably for bargaining purposes, was defined as a 20 to 25 per cent reduction.[54] In October, bargaining began in earnest.

The union went into its negotiations with the NAEL under heavy disadvantages. The barren state of its treasury was public knowledge, as were the increasing number of wage cuts it was forced to resist and the heavy losses suffered by other printing unions. Unemployment, which had become noticeable in the latter part of 1920, had not abated in the industry, and continued to be pronounced throughout the period of negotiations.[55] The employers had in their new secretary an exceptionally skillful exploiter of such advantages. ALA President Philip Bock was a colorful figure and a dedicated trade unionist, but not an astute bargainer, as the 1919 negotiations had already illustrated.

At the end of several offers and counteroffers the ALA leadership proposed that it accept either a flat 10 per cent reduction in wages, or a 12½ per cent reduction for employees receiving more than $30 per week, and a 5 per cent reduction below that. Inasmuch as the average weekly wage for the whole industry was $29 in 1921, probably only some apprentices and a small number of feeders among the organized workmen fell below the $30 figure in the association shops.[56]

In reporting the negotiations to the locals the union leadership had to thread its way carefully between tub thumping and preparing the membership for the possibility of a negotiated reduction in wages. The difficulty of its position is suggested in the following excerpts from a pre-referendum letter to all locals:

The conferees of the Amalgamated Lithographers of America believed that the question of the 44-hour work week should be considered. They were certain of the fact that by waiving the demand for a 44-hour work week they had actually accepted a reduction of 8⅓%.

We can assure the members of the A. L. of A. that the International Conference Committee has used its best efforts towards its continuing our relationship with the National Association of Employing Lithographers, and to make it possible perhaps went beyond the boundary of reasonable compromise; notwithstanding its lenient attitude of compromise the members representing the

N.A.E.L. at the crucial moment ended the conference by refusing to consider any other settlement than the last reduction offered by them.[57]

In Rochester and Buffalo, misunderstandings between international and local union leadership developed, with the result that those locals committed themselves to accepting the 12½ per cent cut before the referendum results on the wage reduction became known. When the referendum results were tabulated and it became apparent by the unanimity of the vote that any wage cut at all would be resisted, the ALA leaders under Phil Bock were forced to taken an impossible bargaining position, which led to personal bitterness between the employers' association and union leadership that continued for nearly a decade. The following quotations illustrate the chasm of mistrust which the unsuccessful negotiations caused:

The NAEL Secretary: The Amalgamated signed an agreement to accept a reduction of 12½% in wages and promptly set out to have it voted down and in doing so disqualified themselves as an organization that can be trusted. It lost them a reputation for square dealing. You can't double-cross people you are dealing with, without loss of prestige. The Amalgamated has so often mis-stated the facts with reference to the 12½% cut, that there is no hope that they will ever tell the truth. The men, who went out on strike in 1922, never have made up what they lost in wages. The large group of men, who kept their word, played square, accepted the reduction and kept their jobs are infinitely better off today.[58]

ALA President Bock: Owing to a depression in business during the early part of the year 1921 and with no regard whatsoever for the equity of the agreement and with no regard to any honest intention to live up to their promise [to introduce the 44-hour work week] which was accepted by the membership of this Association in good faith, the National Association of Employing Lithographers repudiated this agreement, thereby stigmatizing their Association forever as an organization in which no organization nor any individual could place confidence in, as a matter of fact this was the outstanding example of double-dealing that has occurred in recent years in collective bargaining between Employers Associations and their employees.

This repudiation . . . was followed up within a short time by an attempt on the part of the National Association of Employing Lithographers to enforce a reduction in the wages of their employees of 12½%. The larger and more important objective of the enforcement of this wage reduction was to disrupt the Amalgamated Lithographers of America . . .

Nobody could possibly see the necessity of a wage reduction at that time except an element of the National Association of Employing Lithographers who were made blind by an insane urge to increase their profits.[59]

All things considered, one can find little to improve on in the employer tactics. The union leadership had been maneuvered into supporting a wage cut, later rejected by the membership, but at the same time a significant minority of the union membership had been convinced that accepting the wage cut was the wise thing to do. Employers in widely separate locations were coordinating their activities with great effectiveness, and continued to do so after the strike began. In December the union levied a 15 per cent assessment on all members, and employers began attempts to buy off their employees with individual contracts and bonuses. Both tactics had also been used with success in the 1906 strike. The wage cuts were enforced for the first full week in January 1922, and with them came the strike.[60]

There was no possibility of the ALA's paying the constitutional strike benefit of $18 to the 1700 members on strike. Even with the 15 per cent strike assessment, the loss of membership due to the assessment was reportedly greater than the loss due directly to the strike.[61] Benefits for the first weeks were set at $8, and, as the strikers were placed, the benefit was gradually raised to $15 per week. By June the number of men on strike was less than 500, most of these still in New York. With the upturn in business the wage question ceased to be vital for employers. What remained important was the continued operation as open shops, or, in some cases, closed non-union shops. Technically this was in violation of the NAEL rule, a point which caused some irritation to the more violently antiunion members. Such problems tended to be submerged in more fundamental ones, notably, what to do with employers on the opposite fringe who had refused to drop wages by the required 12½ per cent. The decision was to judge "each and every case strictly on its merits," with the result that about twenty member firms were either fined or expelled:

> The Board felt that those members who observed the wage reduction resolution and withstood the strike, suffered very substantial losses which were largely increased because of the nonobservance of the resolution by some of the members and that it is only fair that those who refused to make the sacrifice and incur the loss, should either contribute some amount to the Treasury of the Association or should no longer remain entitled to the benefits of membership.
> This Association must be strong if it is to accomplish the greatest good for its members. Men who can take their medicine and smile are the kind that we want with us.
> The strike year is back of us and we must now look forward to the maintenance of the Open Shop and the upbuilding of our organization. What we want now is added members.[62]

Although the NAEL, like the ALA, was forced to congratulate itself on a decline in membership ("A smaller membership of splendid quality is infinitely better than a large membership that cannot be depended upon," thought one of its members),[63] the NAEL was clearly the winner in the struggle. Table 7 indicates the loss of strength in the union.

Table 7.   ALA membership at year's end, 1921–23: total and selected locals.

Year	Total	New York	Chicago	San Francisco	Boston	Rochester	Buffalo
1921	7351	2811	821	259	206	164	118
1922	5754	2374	672	217	79	71	—
1923[a]	5387	2130	620	168	78	62	—

Source: *ALA Membership Statistics*, internal record of the union.
[a] Although the strike was over by the end of 1922, losses due to the strike also show up in the 1923 figures due to the delay in expulsion of members failing to pay the assessments.

The NAEL was on the whole not dissatisfied with the results of the strike. It was true that success in reducing wages had by no means been complete, but it had ceased to be the major issue. The important point was that the power of the union had been broken in important centers, and control of the labor supply had shifted to the hands of the employers again. In the November employer association convention there were reports of training transferrers in six months, and of the ready conversion of type pressmen to offset work. "Many former feeders, operators and helpers have developed into splendid pressmen," and a word of advice: "When filling these positions look for young intelligent fellows . . . who want to advance and will be helpful in emergencies." [64]

Employer Dominance 1922–1930

The result of the 1922 strike, like the one in 1906, was that the ALA was forced to spend several years recovering strength. The aggressive influence of the NAEL on member shops tended both to keep unionism weak and sources of dissatisfaction at a minimum.

A review of general letters[65] sent to member firms during this period suggests the nature of this influence:

On the open shop: The maintenance of a sound Open Shop situation is a responsibility resting upon the individual employer. You won't keep men out of the Union if you do not give them a square deal and you won't keep them out of the Union if you close your eyes to the fact that the Union Officers are constantly on the job of trying to get men back into their organization.

On individual contracts: You may have some contracts expiring with some of your good workmen. Get them renewed on a fair and equitable basis. If you have disregarded the advisability of making contracts with some of your good men, give some thought to this matter now. Get busy and get your key men under contract. A contract is good protection for an able employee and good protection for employers as well. The man is assured a steady job and you are assured of a man when you most need him.

On fair wages: Wages should be increased for merit. When men are increasing production and giving you good quality of work you should not wait for demands to be made but should make prompt and reasonable increases in wages. Higher wages are being paid in every industry and . . . we must pay as high a wage as the same skill and ability will demand in other fields.

On keeping control of hiring: Our Employment Bureaus [in New York, Chicago and San Francisco] find it very difficult to protect your recently engaged employees if you do not report engagements promptly . . . They can not operate with full benefit to you and others unless you do your part by reporting promptly all engagements, layoffs and discharges.

When you discharge or lay off a man, give the Bureau your reasons for same . . . Good men are scarce and every member should be interested to keep them in member shops. Report promptly so that our Employment Bureaus will know the available men.

On making apprentices: You cannot insure future independence from Union interference more effectively and economically than through the employment and training of apprentices in sufficient number to meet every future manpower requirement. You will have a more loyal and devoted crew and you will ride easy when trouble brews.

On a loyal management group: Foremen cannot carry water on both shoulders. They cannot be loyal to their employer and also to the Union. Foremen occupy important key positions and there should never be any question about their loyalty. Foremen should be non-Union. The rules of this Association require that foremen be NON-UNION.

Employer disunity. While individual employers prospered, the NAEL, like the employees, suffered from the absence of a strong union in the industry. Without continuing evidence of the need for unity, employer members became restive under the restrictions imposed by their organization, and prospective members shied away from the "note feature," the steep initiation fee met with a note payable on demand, the demand being made only if the employer failed to abide by association policy. Some leading employers suggested replacing the note feature with a system of stiff fines on those who failed to turn out enough apprentices, failed to keep their foremen nonunion, and so forth. Others suggested that the open shop could be effectively maintained by eliminating the forty-eight-hour work week requirement and replacing it with rest rooms, libraries, life insurance, and similar items. This prompted one director to comment that if "each group takes a piece out of the frame of the Association and puts nothing in its place there will be left a mud hole where the cellar was." A similar attitude led to the following analysis when the Rochester employers threatened to pull out of the NAEL:

The pay of the Rochester Group as a Group is way under that of the other Groups, and . . . it would be a perfect cinch for the rest of the lithographic industry if they were to go out of the Association, as their help could be gobbled up with the greatest ease without the people who take them on interfering in any way with their own rates.

It seems to me that if this were pretty clearly and definitely pointed out to the Rochester people, and then they were firmly told that if they want to go ahead and go out of the Association and take the consequences, we are prepared to let them go, they will come to their senses mighty quick.[66]

It will be apparent that the NAEL was dealing with a range of problems familiar to any trade union leader, the problems of any organization in which the sources of formal authority stem from below. Members would support it when it delivered the goods, but when there were no goods to deliver, present and prospective members preferred to keep their money and their freedom of action.

Fundamentally, the philosophy of the NAEL reduced itself to two central propositions: the maintenance of the open shop and the forty-eight-hour work week. The strength of these propositions was highlighted in the later 1920's by a reappearance of the historical split in the association between those who saw them as principles which should be defended at all costs, and those who saw them as positions appropriate to the times but subject to change if circumstances changed. The contrast is evident in the attitude toward the forty-eight-hour work week reflected in the excerpts below. Both are from letters between leaders of the organization, the first from the president at that time.

NAEL President: As much as seven years ago we were willing to go on a forty-four-hour week if the Typographical trade did. We didn't go because the Typographical trade only went in part, but it does show that we then were willing to consider the shorter work week. If so, is it unfair now, to give those members [firms] the right to adopt it who want to? We can do that, and hold together on the protection of each other's help, and no doubt gain new members.

I would like to see the Association stay together. I see no other way for it to do so. Do you see a way out?

NAEL Director: Personally I would rather pay twice the dues in a strong organization than half as much in a wishy-washy one, and have no sympathy with the ideas of our self-admitted liberal-minded members who try to camouflage the very definite issue facing us, by talking about "broader scope of activities — better cooperation for the industry — larger membership — local autonomy — new theories of 'brotherly love' for employees" and what-not bunk to camouflage and destroy the only definite flags that we have rallied around for twenty years — to wit: the Open Shop and the question of working hours, or wages. These are the main issues that everybody could be interested in and that we have kept our organization together so successfully.[67]

The 1927 shorter hours struggle. A demand in 1926 from ALA Local 1 in New York that hours be reduced to forty-four in that local's jurisdiction brought matters to a head, and forced the employers' organization (which in August 1926, had changed its name from National Association of Employing Lithographers [NAEL] to Lithographers' National Association [LNA]) to reaffirm or modify its forty-eight-hour work week stand. It chose to reaffirm, and the New York employers, some 20 per cent of total LNA membership, chose to disaffiliate from the LNA rather than face a strike.

The agreement in New York provided for a progressive reduction in hours, dropping to forty-six in April 1927, with a further drop of one hour on January 1 of the two following years. Enboldened by its success, the ALA extended its demands to other major cities, among them Chicago, St. Louis, Baltimore, Philadelphia, Cleveland, Cincinnati, Dayton, and Columbus. In St. Louis it won a clear-cut victory without a strike, which LNA leaders blamed less on employer weakness than on the outstanding abilities of Fred Rose, ALA international vice-president from St. Louis.[68] In other cities the story was far different. In Baltimore the employers drew together into a tightly knit group, determined to eliminate the power of the local, and administered possibly the worst local defeat the ALA ever suffered. Local 18 is a weaker local even today than it was in the mid-1920's. A report by Vice-President Bruck makes clear the employer success in other cities:

Whatever foothold was gained by our Union after the 1922 strike in the Employers Ass'n. houses — was lost again in the 1927 effort for the shorter work week.

Nearly three hundred nonunion men are now employed in twelve Employers Association houses in Chicago. They were partly organized after the 1922 strike, but lost again in the April, 1927 strike.

That same condition holds true in other locals . . . The stronghold of the Employers Association is in this territory and can only be broken by a systematic effort in organization work. As a rule, Presidents of Locals are indifferent or negligent towards organizing. There are some exceptions, but in the majority of Locals — the Officers will not make any effort, believing it to be the duty of the International Office to send an Organizer into the Local.

The Organization has been on the defensive the entire past year. Very little could be accomplished except to keep the membership intact. When it was decided to retreat last June, I endeavored to place the striking members (of the Cincinnati, Dayton and Columbus locals) as quickly as possible — so as to regain a foothold in the various struck shops and reduce assessments as well. This has been accomplished to a great extent as only about forty-three members are on benefit list in the three locals.

While the situation in Baltimore is still very unfortunate we must look forward to the time when the strike assessments will have to be taken off entirely [i.e., discontinue strike benefits to the men still out] . . . Organizing is difficult under any circumstances, but more so when a strike assessment is being collected.

For the information of the International Council I wish to state that in the year 1928 — I spent 330 days on the road.[69]

Yet the employer victory in 1927 was a far less impressive one than that in 1922, just as that one had been less impressive than the one in the great strike of 1906. Hours had been broken in New York and St. Louis, and in commercial shops in almost every city. Although the LNA had most of the larger color houses as members, it now had less than 20 per cent of all lithographic firms,[70] and well over half the ALA membership was in non-association shops. Moreover, the defection of the New York employers left the LNA so short of funds that a number of firms had to underwrite the deficit in 1929, and it became increasingly clear that an organization which restricted membership to open-shop firms with a forty-eight-hour work week was more apt to disappear than grow.[71] In a resolution that must have caused much mirth among the "liberal employers" and in the union, the 1929 LNA convention bowed to the inevitable. Beginning with five paragraphs about the work of the world, our complex industrial life, and the fact that things do not create themselves, the sixth and operative paragraph ended, "members may establish a work week of less than forty-eight hours." [72]

The proximate cause of the change in policy, accomplished with far less grace than the change from fifty-four hours to forty-eight hours in 1911, was the defection of the New York group. As the LNA secretary pointed out, "the Association has not recommended a reduction in hours. We have, however, made it possible for open shops that are operating less than forty-eight hours to become members of this Association. That is the net of the action taken." Sentiment was strong against the New York group in some parts of the membership, its agreement to the hours reduction being described by one director as doing "incalculable damage to the trade . . . it would have been a short and sweet fight with them if they had called the

Union's bluff." [73] This was not entirely hindsight, for the outcome of the Baltimore, Miami Valley, and Chicago disputes was still in doubt. In any event it proved accurate, for in December the New York local agreed not to enforce the hours reduction from forty-six to forty-five on January 1, 1928, in view of the "high cost of production, the depressed conditions of business, [and] the amount of unemployment now existing." [74] An agreement *not* to reduce hours due to substantial unemployment suggests that the union was not bargaining from a position of strength, a suggestion supported by the final paragraph of the *Lithographers Journal* report of the agreement:

> It is the belief of many that Local No. 1's action will have a widespread beneficial effect. It will prove once more that the ALA is inclined to refrain from extreme measures to the detriment of the industry. It will also make the continued support of lockouts and strikes in Baltimore, Dayton, Columbus and Cincinnati more certain.

In mid-1929 all strike benefits were discontinued in Baltimore and other cities where some of the unemployed members were still officially classed as strikers, and in mid-July ALA members were free of assessments for the first time in over four years. During the two years, 1927–1928, the ALA had spent over one million dollars and lost four hundred members in an only partly successful attempt to gain an hours reduction that other printing unions in large part had already attained.

Chapter **VII** ✆

The System during the Thirties and the War Period

The thirties were lean years for many unions, the ALA among them. Lithographic employers had contained the union effectively during the previous decades, and the union was internally weakened by unevenness in the degree of organization among the several crafts.

Pressmen were well organized, but the feeders were not, and the potential harm to employers of a strike was thereby lessened. Councillor (later president) Kennedy, himself a feeder, made this point forcefully in the 1927 convention:

> In Baltimore, they regretted from the day it [the strike] started [that] some organizing hadn't been done with the tin feeders previous to that time.
>
> I am pointing out to you that if they don't in some way try to organize the tin feeders and pressmen in back of them — you call them hunkies and other names — but they are there and if you think they are blind and can't see, you are fooled. They know how the press is being operated.
>
> It ought to be a lesson to us, and we should not overlook the fact that the feeder end on the automatic and the hand feeder, tin or paper, is an important part of the industry and especially when you are out on the street, walking it for months and you shouldn't overlook the lesson Baltimore is teaching us now.[1]

Substantially the same problem arose with the artists, engravers, and designers. The nature of their work lent itself to subcontracting, and enterprising craftsmen would set up an operation of their own. Such operations inevitably undercut certain production costs in a regular lithographic firm, and they were very difficult to organize or keep organized, since often the only worker was the employer himself, who would hire friends on a part-time or temporary basis to handle additional work. And in times of strike, trade shops could be fully as dangerous as unorganized pressmen:

> *Brother Hoesterey:* The trade shop evil is the undoing of the engraver and the artist in the city of Rochester. We made a mistake a few years ago and

allowed one of our members, an officer, to open a trade shop under the Constitution, well within the law. When the trouble came on, he wasn't with us. He was our financial secretary and we have never been able to get his books yet. That is aside from the issue.[2]

Trade shops fulfilled a necessary function in the industry, but they permitted an employer to dispense with the services of regular artists, engravers, or designers, and then these workers would set up their own trade shops to further jeopardize the wage levels of regularly employed craftsmen in these trades. The onset of photo-mechanical platemaking methods and their adoption by the trade shops eliminated the need for hand transferring images to the lithographic plate. The seriousness of this problem for the transferrer in 1927 is suggested by the following figures: out of 131 cameras in litho plants and trade shops of which the union was aware, only 56 were operated by union men, and of 120 "transfer machines" (photo-composing machines), only 66 were union-operated.[3] The pressman could greatly assist in reducing this problem by refusing to run plates made in nonunion shops, but this was asking for more than was possible. The trade shop problem was very real for the union as the nation started the slide into the great depression. Like the problem of the feeders, it gave evidence that the ALA was still an incomplete amalgamation of crafts, rather than a union of lithographers.

Depression: 1930–1933. In 1930 the employment slide was just getting well under way. The employment index in book and job printing had dropped from a December 1929 high of 110, to 100 in June 1930; yet it was to drop 33 points more to 67, before the low point was hit in April 1933.[4] Unemployment among ALA members, which was in the neighborhood of 10 per cent in 1930, had risen to 25 per cent by 1932, with 47 per cent of the journeymen members on part time.[5] Had apprentice members been included, the unemployment percentage would almost certainly have been higher.

There is an element of pathos in the discussion of unemployment in the 1930 convention proceedings. The delegates had heard that the 1927 strikes had so depleted their treasury that money had to be borrowed from the mortuary fund (death benefit fund), and they had sat through the State of Association reports that recorded a general decline in the strength of the locals and an almost uninterrupted series of lost strikes.

No one suggested that socialism was the answer. Even the Socialist editor of the journal, Carl Halbmeier, contented himself with references to "Wall Street laying off men by the thousands," and to the minimum concessions made by President Hoover and the government "to prevent labor from getting a revolutionary spirit." Hoey of the feeders suggested demanding a shorter work day from the employers, and with a sublime indifference to the recent events in Baltimore, the Miami Valley, and Chicago, he said the organization should take on "a little bit more courage" and be "a little bit more daring." For, said Hoey, "It seems to me in the past we have been too timid." Al Castro ran Local 1. Suggestions for daring a bit more had no appeal for him, nor did the work-sharing devices which several members suggested. He pointed out that as a practical matter employers would not tolerate a rotating workforce. His suggestion was for the president to send letters to open shops and ask for cooperation.

Councillor Rogers summed up another viewpoint when he said that the ALA was "founded in one word, that is 'Charity.' We know the bosses are not going to melt into our idea of things; we cannot force them; the thing is whether you are willing to dig down into your pocket and help your brother." Vice-President Kelly agreed. "Under the present industrial system there is no remedy or solution . . . All we can hope to do as an organization is to alleviate the suffering."

In no other convention of the lithographers has a counsel of despair been so openly offered — and so generally accepted. Yet in this respect the lithographers were little different from their brothers in the rest of the American labor movement:

> In some ways the most surprising phenomenon of the depression years was this apathetic attitude on the part of industrial workers while the unemployment figures steadily mounted and the bread lines lengthened. There was no suggestion of revolt against an economic system that had let them down so badly. There was no parallel to the ugly railway strikes of 1877 or Debs' Rebellion in 1894. In Park Avenue drawing rooms and the offices of Wall Street brokers, there was a great deal of talk of "the coming revolution," but the unemployed themselves were too discouraged and too spiritless to be interested.[6]

The employers, and particularly the association employers, could find some satisfaction in the weakened state of the union. The *National Lithographer,* which for many years had been the mouthpiece of the employers' association,[7] had stood aghast at the presumption of the union in demanding the shorter work week in 1927.[8] Had not the Lithographers National Association (formerly NAEL) said it would oppose this? And "when the LNA goes into a war, it goes in to win." The LNA, however, had been winning its wars by smaller margins and with larger defections each time. The 1927 "war" was barely a victory, and it was at the cost of substantially all the New York membership.

The LNA leadership of the fight against shorter hours unquestionably had enjoyed a measure of success, as Table 8 illustrates.

Table 8. Hours of work in the printing industry, 1929.

Industry	40 hours and under (in per cent)	Over 40 but under 45 (in per cent)	45 to 48 hours (in per cent)	Over 48 hours (in per cent)	Total
Lithographing	1	4.5	93	1.5	100
Book and job printing	1	45.0	46	6.0	100
Photo-engraving	—	72.5	25	2.5	100

Source: *1929 Census of Manufactures*, vol. I, pp. 54–55.

Yet this leadership had so decimated the LNA's ranks and so strapped it financially that it was forced to modify its hours stand, a decision which did not in fact relieve the financial problem. As the depression deepened, many firms were forced to cut away all nonessentials, and even a 25 per cent reduction in dues did not save the association from a steady loss of members. Increasingly the content of communications to member firms focused on product market rather than labor market problems, as the former became the central and most pressing concern of lithographic employers.

The National Recovery Administration

The NRA is one of the more remarkable phases of American political and economic history.[9] It proposed to give the benefits of price-fixing to all businessmen, in order to preserve the free enterprise system. Many of the codes contained obvious featherbedding provisions, yet were written by businessmen. It was the national salvation when initiated, and the national runaround eighteen months later.

In 1934 President Roosevelt summarized the purposes of the NRA:

We seek the definite end of preventing combinations in furtherance of monopoly and in restraint of trade, while at the same time we seek to prevent ruinous rivalries within industrial groups which, in many cases, resemble the gang wars of the underworld and in which the real victim, in every case, is the public itself.

Under the authority of this Congress, we have brought the component parts of each industry together around a common table, just as we have brought problems affecting labor to a common meeting ground.

Though the machinery, hurriedly devised, may need readjustment from time to time, nevertheless, I think you will agree with me that we have created a permanent feature of our modernized industrial structure and that it will continue under the supervision but not the arbitrary dictation of government itself.[10]

In April 1933, according to Secretary of Labor Perkins, Roosevelt's mind was "as innocent as a child's of any such program as NRA"; a year later more than 200 code authorities had been established for industries employ-over 15,000,000 people; and in another year the program was emasculated by the Supreme Court's decision in the *Schechter* case. By January 1936, the NRA was history. But in an eighteen-month period, 556 codes covering 22,600,000 employees for industries ranging from "pickle packing" to "legitimate full length drama," had been promulgated, with 201 supplementary codes to cope with segments of the more inclusive industries. The 700 code authorities actually established, typically the boards of directors of the dominant trade association in each code-defined industry, covered industries employing from 1.2 million people (the trucking industry) down to eight industries for which the code coverage was less than 150 people.[11]

The Lithographic Code. The NRA forced the lithographic employers to reconsider the relative merits of conflict versus cooperation with the union. No one knew what the practical effect of the codes would be, but few employers quarreled with their expressed purpose — the stabilizing of prices to protect "fair" employers from "unfair" competition. Particularly was this so in the lithographic industry, where the number of employers competing in the same market was small enough to make national price uniformity seem attainable in principle, yet too large for it ever to be realized in practice. It was reasonable to hope that the legal and moral sanctions available through governmental sponsorship might be enough to do the trick.

The LNA, however, was faced with a dilemma. It was the natural choice for "code authority" of the lithographic industry, but its restriction of membership to open-shop employers excluded it from consideration. A members' bulletin stated the problem with a tactful clarity that was the trademark of the secretary of the LNA:

> Industries are going to be permitted to set up their own machinery for fixing the hours to be worked, the reasonable rate of wages that should be paid, and to provide reasonable control over competition to eliminate unfair methods, cutthroat prices and other undesirable practices.
>
> Whether our Association can be geared to meet the requirements of the new situation will have to be discussed. Whether or not lithographers want to organize a new association, permitting our members to support our Open Shop policies, is a question to be considered. Certain it is that if lithographers do not join in a cooperative effort to eliminate the many abuses which have grown up, and to bring about a stabilization of prices that will get back the money spent in production, and eventually return something to the stockholders, our industry is doomed to go down, and down, and down.[12]

The open-shop principle, which had been so long and so closely held, went by the board in a single convention, and the LNA was prepared to lead the way to "controlled competition" in the lithographic industry. Most lithographers were betting on it, and in a space of three months membership skyrocketed. In the month of August 1933 alone, more than one hundred membership applications were received.

The first danger to be avoided was a single graphic arts code for the whole printing industry. The letterpress associations were working hard to bring all printing firms under a single graphic arts code, and their much larger size made such an outcome probable. The NAEL secretary believed that the lithographic industry was "fighting for its life," and in this fight saw the ALA as an ally rather than an enemy. The ALA was even more interested than the NAEL in securing a separate lithographic code. President Berry of the pressmen had seen the potential for using a unified code to settle the jurisdictional dispute with the ALA to his own satisfaction. The ALA's new president, Andrew Kennedy, had sensed this danger, but had anticipated that no cooperation would be possible with the lithographic employers. But he and the LNA secretary found common grounds for understanding almost immediately. President Kennedy noted with pleasure that "On the whole and rather surprisingly, our employers through the Lithographers' National Association were fair enough to bargain with us in an amicable spirit and as is evident, we secured the best modifications that were passed for any industry by the NRA."[13] Eventually the modifications written into this temporary code for the lithographic code were eliminated, but the cooperation between the ALA and the LNA continued and was directly responsible for making the lithographic segment of the graphic arts industry one of the four autonomous administrative units under the Graphic Arts Code.[14]

The effect of the code. The employment provisions of the code became the standard for the industry, so much so that in 1936 an amendment to the ALA constitution was sent to referendum to bring the constitution "in line with the code regulations on overtime and apprentices since the code rules on these two questions are now the accepted rules for the industry."[15] Most important was the change in hours. In 1930 the industry had moved away from a forty-eight-hour work week but had not yet gone to forty-four. The NRA had the effect of standardizing the hours at forty, with very few exceptions.

The effect on unemployment among union members is evident in Table 9.

Table 9.　*Unemployment and part-time employment among
ALA journeymen members.*

Date	Per cent unemployed	Per cent part time
June 30, 1933	24.4	38.9
December 31, 1933	17.7	24.4
June 30, 1934	12.3	21.6
September 30, 1934	12.3	19.2

Union figures also indicated that the method of compensating for the hours reduction had provided a 10 per cent increase in hourly earnings and a 7 per cent decrease in weekly earnings. The ALA found much to complain of in the operation of the code provisions, but it much preferred them to no provisions at all. On May 28, 1935, the day following the invalidation of the enforcement provisions of the NRA in the *Schechter* case, Kennedy sent a letter to all locals to strike immediately and without reference to the international any employer who attempted to ignore the hours or overtime provisions of the code. His statement of the problem was illuminating: "the Supreme Court decision places the task of maintaining these conditions on our organization." [16]

The actual risk of having to fight a large number of strikes was not great. So long as the LNA shared with the ALA the conviction that the NRA code provisions should continue to be the basis of the terms and conditions of service in the industry, there was no real danger that the ALA would be faced with a large number of recalcitrant employers. In point of fact, during the period 1935–1937, the ALA was carrying out the function which had been given to the employers' association during the period 1906–1917, the policing of employers to ensure that they conformed to the standard conditions which had been negotiated during the period of the NRA.

Collective Agreements during the 1930's

With the passage of the Wagner Act in 1935, the importance of written agreements was raised substantially. During the 1920's there had been little emphasis on written agreements, primarily because LNA shops refused to sign them, but also because the ALA preferred to enforce its constitution where it had substantial bargaining power.

When an employer wanted to use the union label, the working rules would be formalized in a brief document:

Witnesseth, that the said party of the first part, in consideration of the use and privileges of the Amalgamated Lithographers of America's Union label, owned and controlled by the said party of the second part, hereby agree to employ in the different departments of which the party of the second part claim jurisdiction over none but members of the union of the party of the second part and to comply with the adopted rules and regulations of the union represented.

To see that all work contracted by _____ is produced by union labor, as heretofore provided, not to use the said union label upon anything but the strict production of such union labor, and not to loan said union label, except by permission of the party of the second part. Any violation of this agreement shall make it null and void; and the further use of the Amalgamated Lithographers of America's Union label shall be without warrant and illegal.[17]

The right to the label was important on some types of printing, but particularly important to trade shops which supplied lithographers with engravings, plates, and so forth. Even these agreements, however, did not have a specified time period, with the result that if the firm was no longer in need of the label, the ability of the union to enforce the standard conditions was in no way strengthened by the agreement.

When Andrew Kennedy was elected ALA president in 1930, the union attitude toward agreements changed; so also did the economic situation. President Kennedy, together with the other leaders of the union, spent much of the following two years traveling from city to city, trying to control wage cuts by employers, securing their agreement on work-sharing devices, and dealing with other problems that would have been eliminated or much reduced had written agreements been in effect. The experience convinced the ALA leaders that time spent on contacting employers individually each time a wage cut was threatened could far better be spent on annual negotiations, and the time saved used for organizing.

The first opportunity for negotiations between the union and a group of employers occurred in 1933 in the NRA code discussions, when the initial effort of the employers to get a code approved without reference to the union failed, and they recognized that they would either have to work with the union or face a long delay in securing approval of a code for the industry. This was bargaining, but with a much greater degree of government intervention than was normal.

The ALA fought vigorously to have the hours reduced to thirty-five with no loss in wages, and, though not successful in this, it was able to secure this provision in the temporary code:

Pending the approval of the Basic Code of Fair Competition of the Lithographic Industry, present hourly rates shall be increased so that the weekly payments for forty hours shall not be less than the payments now made for present basic work weeks.[18]

This provision meant a full makeup in weekly wages, something which no other printing code contained. The NRA disallowed this provision, and, because keeping it would have in any event meant that many lithographic firms would have sought coverage under the more relaxed printing codes, the ALA was not in a position to object.

Shortly after the independent lithographic code authority was approved, the ALA and the employers cooperated again in securing an independent lithographic Industrial Relations Committee, whose purpose was to secure compliance with the labor provisions of the code.[19]

The Industrial Relations Committee had no legal grant of power to enforce its decisions, although it was supposed to deal with violations by uncooperative employers. Yet even here President Kennedy found progress of a sort possible:

> We have agitated for the establishment of minimum wage scales arrived at through collective bargaining among individual employers, local groups of employers and within the Industrial Relations Committee, the result being that we are certain that the subject of minimum wage scales will receive considerable discussion at the employers' convention and that we will have many supporters among the employers for our position on the question.[20]

It was at this point that the employers chose to select a young lawyer who had been retained as counsel to the code authority, and to assign him the job of enforcing the labor provisions of the code. This was Benjamin Robinson, who later became the devoted friend of Andrew Kennedy, and a mighty warrior for the ALA.

The activities of the Industrial Relations Committee continued after the *Schechter* case had emasculated the NRA. So did the cooperation between the union and the LNA:

> When the Supreme Court declared the National Industrial Recovery Act unconstitutional, we were able to secure cooperation from the Lithographers' National Association, to the extent that they agreed that despite outlawing of the codes, the employers' association would cooperate with the union in the continuation of labor conditions as were provided under the code. It must be conceded at this point that the officers of the Lithographers' National Association have cooperated where attempts were made to increase the maximum hours, or violate some other labor condition which was part of the code.[21]

The NRA had a substantial impact on collective bargaining in the lithographic industry. The enforcement procedures depended fundamentally on "a great spontaneous cooperation";[22] if this cooperation was not forthcoming, then the importance of the legal setting was not substantial. But in the lithographic industry a strong union and a substantial group of employers were interested in cooperation. It was a good deal short of the spontaneous type President Roosevelt had in mind, but it established a working relationship between the leaders of the union and the leaders of the LNA that had been completely severed by the 1922 strike. Both the ALA and at least some part of the LNA leadership looked forward to the time when wage costs could be again stabilized in the industry by means of a national contract. The LNA secretary addressed the LNA convention in 1937 on the subject of minimum wages, detailing the unjustifiable spread between low and high hourly rates for several classifications (35 cents to $1.75, 75 cents to $2.87, and so forth). "Subnormal wages cannot continue to exist in this industry," he continued. "Maybe some of you will be smart enough to get away with it . . . but the whole wage movement is against you." [23]

A significant portion of his address to the convention was devoted to the presentation of facts showing that a national agreement would be of value to the employers. The wage differentials cited above were quite naturally one of the best arguments in favor of such an agreement to those companies whose wages were high.

It will be useful to comment somewhat further on this total reversal of the LNA attitude toward the union in the space of five years.

The ALA during this period enjoyed outstanding leadership. Fred Rose in the midwest, Robert Bruck in the central region, and Andrew Kennedy and Albert Castro on the east coast provided the union with men whose ability to lead was not questioned by the membership, and whose awareness of industrial problems was evident to the employers. One of these problems was the expansion of the lithographic process into shops which were not in the LNA and would not join that organization. The ALA worked actively to gain control of these shops and was reasonably successful in doing so. The LNA had to depend on the ALA to do what historically had been the function of the LNA, the maintenance of uniform working conditions.

Finally, the ALA was the one union that could and would organize all the valuable workers in the industry. When the Wagner Act made the nonlithographic workers count for something, the ALA expanded its jurisdiction to include them. Dealing with one union was preferable to dealing with several; dealing with an AFL union was preferable to dealing with the CIO; and dealing with a known quantity was preferable to dealing with the unknown. On all counts, the ALA was the least of evils and in some ways a positive good.

The 1937 Agreement. As early as 1934 there had been open discussion of a new national agreement for the lithographic industry.[24] The attitude of many employers was not favorable for such an agreement, because they preferred to continue on a nonunion basis or with the informal relationship that had become the standard following the 1922 strike. Apparently because of the validation of the Wagner Act, the LNA determined that it would be wise to sign a national agreement with the ALA, and it successfully concluded negotiations in June 1937.[25] Basically, the agreement put the working conditions into contract language in respect to hours, overtime, and apprentice provisions that had been established during the NRA period. In addition, one week's vacation after one year's service was included, and a clause on wages providing for some increases. Like earlier national agreements, there was no national wage scale for all classifications in the agreement, but there was a strong emphasis on evening out interregional wage differentials. This portion of the contract is given in condensed form:

Section III. Ten per cent increase to workers drawing less than 70 cents per hour, 5 per cent increase to workers drawing more.

Further increases as necessary to raise the employee to the rate received on July 1, 1929, but not to exceed 12½ per cent or the 1929 minimum for that classification in that area.

Section IV. Supplemental agreements between individual firms and their employees may be desirable. In making such agreements "the parties declare that it is their intention to cause throughout the country conditions of employment as nearly uniform as may be practicable, taking into consideration differences in local conditions which in fairness to all parties would cause justifiable variations."

The agreement was accepted in referendum by the ALA membership, and approved by the LNA board of directors. But out of the approximately three hundred firms in the LNA, only thirty actually signed the agreement. It had been recognized that under the terms of the Wagner Act it would be necessary to have each firm sign the agreement since some of the association members were not organized by the ALA. But it was not anticipated by either the union or the LNA leadership that only 10 per cent of the firms would support the action of their own association. President Kennedy avoided making any statement critical of the LNA, but Benjamin Robinson, by that time a close friend and regular consultant to the ALA president, was less generous. He defined the failure of the national agreement as the failure of the LNA to sell its own membership on the desirability of such an agreement.[26] The ALA was disgusted with the waste of time and effort spent on the negotiations, and the international council agreed in late 1938 that "no agreements should be negotiated with a national group of employers until the employers can convince us that they are able to, and do, speak for all the members of their association."[27] With the wording changed only slightly, this was the statement that the LNA (then the NAEL) had made to the union officers before the 1906 strike, the employers insisting that they would deal only with a union committee with full power.

The LNA had lost its power. It actually had lost it in 1927 when it was unable to present a united front on the question of hours reduction. The weakness of the ALA resulting from the 1927 strikes, internal dissension, and the depression had not given the association any test of its internal strength prior to 1933, when the NRA had offered a measure of protection. This protection had been continued by the desire of the ALA to work in harmony with the employers during the post-NRA period. But the inability of the association to commit its members to an agreement that was scarcely more than the writing of the *status quo* into a formal contract was so obvious a sign that the association was through as the representative of employers that it ceased even to work actively in the field.

In September 1937, the eastern lithographers' association (the New York City employers' association) appointed a labor committee. In reporting the event the *National Lithographer* added that "The association recognizes that the subject of labor is national in scope and, therefore, primarily within the province of the national association."[28] But outside the pages of the *National Lithographer,* the subject of labor had ceased to be national in scope, except insofar as the ALA was able to make it so.

Development of the Union

The inability of the LNA to act as a unifying force in the industry was not an encouraging development for the ALA. The burden of preserving the separateness of the industry fell increasingly to the union, and the necessity of doing so stemmed from its dual position in the eyes of the other printing unions. The charge of dual unionism has been a headache for union leaders, but its effect on the union membership is not always an unalloyed evil. There have always been strong pressures for democracy within the union. The active presence of other unions interested in representing the ALA membership has also forced the amalgamated to concern itself directly and continually with attaching its members closely to it, and with giving the minimum amount of reason for discontent or for questioning whether one of the other unions could do a better job. These pressures helped shape the form of government in the union and led directly to the switch from craft to regional representation on the council. Even by the late 1920's, members were no longer satisfied that a fellow craftsman from any location might represent them. "My local," rather than "my branch of the trade" had become the primary bond of union allegiance. To a new generation of members, the top hats of the artists were a legend, and the fulminations of Hoey against the indifference shown the feeders barely a memory. Craftsmen in dying branches were being retrained and reemployed in the new and growing skill classifications, and feeders were secure in their assured preference when apprentice pressmen were chosen. The relevant problems were those that the action of one local could cause for another.

The most alarming of these problems was the loss of work when wages of one local were too far out of line with those of others. Until the late 1930's, interlocal wage differentials were not a serious problem for the ALA. For example, from the evidence available it would appear that the standard deviation of wages for pressmen was 5 per cent of the average wage in 1923 and 15 per cent in 1947, the first year of the 1940's for which statistics are available. In 1923 Local 1 pressmen were 17 per cent above the national average; in 1947 this figure had jumped to 26 per cent.[29] Both these calculations are based on incomplete data, yet it is worth noting that 1923 was the year following a bitter strike in which a number of locals had caved in, and 1947 only shortly followed the long period of wartime wage controls. If members were to protect their employment opportunities, it was more important to control interregional wage differentials than intercraft ones. The issues which brought fire to the ALA conventions no longer involved craft differences; they were issues revolving around interlocal competition.

The general workers. The necessity for preserving the separateness of the industry also led the ALA into another significant break with the past — opening its membership to unskilled workers. Some employers specifically wanted the ALA to organize all their workers, and offered no hindrance and occasional assistance when the ALA requested bargaining rights for the whole plant.[30] President Kennedy for his part was firmly convinced of the wisdom of taking the nonlithographic workers into the union. When the Rochester local told him that the CIO was trying to sign up men in one of the litho shops, and was getting them because its initiation fees and dues structure were so much lower, President Kennedy instructed the local officer to make his price competitive and "go out and grab 'em." [31] At the end of 1936 there were no general workers in the union; one year later there were 3000, or about 25 per cent of the membership.

Although numerous locals consistently refused to accept general workers into membership, the general worker group continued to grow in numerical strength throughout the war and early postwar period. As late as 1947 the general worker had strong support from an important segment of the leadership of the international. Pacific Region Councillor Brandenburg wrote an article in the *Lithographers Journal* on "Why General Workers Must Be Organized," President Riehl's motto was "double the membership of the ALA," and much discussion in the journal, the convention proceedings and the general letters to the membership revolved around the strength that industrial organization gave to a union.

The year 1947 marks the high point of this development; since that time there has been a steady decline in the importance of the general workers in the ALA. There are several reasons for this. The general worker was accepted somewhat uneasily into the ALA. Most of these people were unskilled women workers, employed in the finishing department and therefore not closely associated with the actual lithographic process. For a time the phrase "general worker" was displaced by "finishing and maintenance department help" to express the fact that some skilled mechanics were also included in the catch-all category. But "general worker" was descriptive of what ALA craftsmen thought of their new brothers, and the name stuck.

A second quality of the general workers came as something of a shock to the leadership — they did not mind being on strike. The ALA has always paid strike benefits, and in general has tried to have them reflect the needs of the workers involved. When in Philadelphia, then in Buffalo, and soon after in Poughkeepsie the union discovered that at the level of strike benefits paid by the union the general workers were willing to stay out rather than compromise one iota on the demands which affected them, the advantage of having general workers in the bargaining unit came to be viewed in a different light. Councillor Brandenburg had said in his 1947 journal article that the general workers had "strengthened our Local and in our recent negotiations this helped the Negotiating Committee by showing that in industrial organization we have strength." [32] Presidents Luke of Buffalo, Willis of Boston, Swayduck of New York, and O'Neill of

Poughkeepsie were by 1954 able to offer considerable evidence for the proposition that general workers could also make a hash of reasonable bargaining strategy.

In fact, the actual decision to move away from organizing general workers was a part of a broader shift in the ALA's approach to the jurisdictional dispute with the printing pressmen.

Organizing activity of the printing pressmen's union. It must be conceded that the printing pressmen had not shown much interest in taking possession of the offset jurisdiction awarded to them in 1915. Other than vigorous presentations to AFL conventions, NRA administrators, National Labor Relations Board Hearing officers, and similar functionaries, little seemed to happen in the way of organizing. There appear to be two important reasons for the inactivity.

In the 1920's and early 1930's there were relatively few combination plants. President Berry's quite legitimate fears that lithography would cut seriously into the commercial printing markets proved true only in part, and the market went to the firms with the new process (the offset press), not the new process to the firms with the market. The offset press was not sufficiently simple in its operation to be introduced without skilled personnel, and it did not provide large enough savings for enough products to endanger seriously the commercial printer if other firms adopted it. As a direct result of this, the International Printing Pressmen and Assistants Union (IPP&AU) had very few offset pressmen, and there was little pressure on the leadership from this source to get out and organize the competition.

The second point involves relative wage levels. The IPP&AU was not equipped to train apprentices or press assistants to run offset presses. Its trade school operated on the assumption that a relief pressman had nine tenths of the skill required to operate an offset press, and that the additional training time necessary was only six weeks. Quite apart from the dubious validity of the assumption, it tended to draw men who were already earning journeyman wages. Few jobs on an offset press within their qualifications would have given them anything but a reduction in wages. The result was that letterpress firms, when they did go into offset, would in many cases be forced to appeal to the ALA to supply them with qualified workers. The fact that relations were cordial between many of the local pressmen and lithographic unions would often permit this to happen with little or no interunion friction. President Berry, with the help of the barber, the fireman and the patternmaker, had won for the IPP&AU the right to represent offset pressmen, but the ALA continued to be the only union which could supply them. Though this was true in many instances, none was so incisively described as the situation in Minneapolis, in a letter in 1945 from the general manager of the local trade association to an official of the IPP&AU:

Our position is briefly this: this association has, in Minneapolis, Minnesota, attempted to cooperate with International Printing Pressmen & Assistants' Union with respect to the operation of offset equipment by members of that organization. In every case with which we have had direct contact, it has been impossible for International Printing Pressmen & Assistants' Union or its locals to cover their jurisdiction where such jurisdiction was recognized or invited by employers.

Within the past four months Argus Publishing Company tried to get a man from Local 20 to run their 17 x 22 Webbendorfer offset press, and they waited for three months during which time the press stood idle, and finally got disgusted with promises of officials of Local 20 to "try to get a man here from California," "believe we can bring a man up from Iowa," "the International thinks they have got one located in Wisconsin," (or maybe it was Indiana) until finally the officials of Local 20 threw up their hands and said to forget about it because the International didn't have any and they relinquished their claims. The Argus Publishing Company then called the Lithographers' Union, got a man who is operating the press with good results.

Four and one-half years ago, at a time when the manpower situation was not so stringent as it is now, Colwell Press put in a 17 x 22 Webbendorfer offset press and called upon Eddie O'Neil, who was then the business representative of Local 20, to furnish a man. Eddie O'Neil furnished a man who was unable to start the press, and that fact was not discovered until he had been on the job for half a day. He then stated, in answer to questions as to why he was not operating the press, that he was a barber by trade and he had during a vacation the year before observed his brother-in-law run such a press in Rockford, Illinois, and was "sure he could run it if somebody would show him how to start it." [33]

Because President Berry had used Jacob Fischer of the barbers' union on the adjustment committee thirty years before, humorists suggested that George was now paying off old debts. The IPP&AU had not yet become a factor in the lithographic industry.

Union-Management Relations during the War

The failure of the 1937 national negotiations between the ALA and the LNA had given additional impetus to the formation of local employer bargaining groups. By 1941 group contracts were being negotiated in perhaps half of the larger cities, in most cases with locally organized groups of employers.

The negotiations which occurred during the war period were limited in scope. Reports of officers and representatives during this period are limited largely to accounts of getting the maximum allowed by the war labor board, appeals for special consideration, and in other ways spending time before the government agencies to secure concession for the membership. By and large, the period 1941–1945 was an extension of the period of labor-management cooperation in the industry which had begun in 1933 with the NRA. Negotiations under wage controls tended to make the government the goat when differences arose, and the union was still led at the top by men who were firmly committed to the idea of working together with management, men who had been leaders throughout the depression and had learned to place a high value on secure and steady employment.

The Joint Lithographic Advisory Council. The National Recovery Administration had given the union and the LNA an opportunity to work closely together. The formation of the industrial relations committee had permitted this cooperation to continue after the end of the NRA, and the 1937 contract had given some formal continuity to the working relationship. Toward the end of the war a new effort was made to provide a continuing forum in which problems of mutual concern could be discussed without the necessity of choosing words as carefully as would be required in a negotiating session. Largely through the efforts of Benjamin Robinson such a body was established, the Joint Lithographic Advisory Council (JLAC). No verbatim minutes were kept, and throughout its rather short existence the JLAC maintained an atmosphere of informality which considerably furthered its ability to serve the purpose intended.

The decision to establish the JLAC was taken by the LNA and the amalgamated, with no mention of other employer groups participating. Other employer associations thought that they ought to be represented, and that the LNA had only a small claim to the position. It is noteworthy that these convictions were taken to the union, so that the union would see to it that employers got fair treatment from each other.

These letters of complaint from employer groups provide dramatic evidence of Andrew Kennedy's success in reshaping the goals of the union, and casting them in terms of the industry rather than of the labor movement. The JLAC itself is an impressive testimony of his success in this effort. Not only was the union a partner of the employers in their planning for the postwar period, but it was the organization around which employer groups were able to gather when they could not join together by themselves.

Benjamin Robinson wrote to the ALA officers in September 1944:

The Amalgamated has achieved its main objective, namely national recognition of the Amalgamated by the employers and management's acceptance of the Amalgamated as an integral part of the industry. We must emphasize this theme.

On the other hand, there is much disunity in management. There is a striving for power between the two national groups [LNA and NAPL], as well as between either or both national groups and local organizations. It is our job to weld them into a more cohesive group. It will be more to our benefit in the long run to have a strong national group than to have the knifing by many small groups.[34]

This attitude may be contrasted with that illustrated by a clause successfully demanded by the ALA in a 1927 contract:

[The Company] shall not during the life of this contract enter into any association or combination hostile to the Amalgamated Lithographers of America, nor shall it at any time render assistance to such hostile combination or association by suspension of lithographic work or by any other act calculated to defeat, impede, or interfere with the policies or activities of [the union].[35]

The ALA was no longer on the outside looking in at a strong employer organization, writing its contracts to help weaken it. It was trying to be a mature, conservative organization, looking toward the time when it would be called upon to fulfill the role of industrial statesmanship which President Kennedy had foreseen, and which the turn of events had now brought so close to reality. Benjamin Robinson had much to do with the original setting-up of the JLAC, and it is useful to quote further from his memorandum to the ALA officers, keeping in mind that at this time his influence in the ALA was very nearly at its peak:

This is an Industry Council. It is interested in every phase of this business: (1) selling (2) production (3) distribution and (4) competition. Over all these specific phases, the Council will be motivated by its goal — a stable, prosperous industry . . .

The basic problem concerns the growth of this industry and the manpower question in the industry. Related to these are the problems of the Union's apprentice ratios, work classifications and other items. It is obvious that the employer will raise as issues those things which call for modification of the Union's restrictive provisions. Now it is important for our representatives to understand that this is the only possible approach and we must not resist such an approach by the employers. We should be perfectly willing to recognize that in a balanced industry certain restrictions would fall, and that certain other restrictions made sometime ago do not conform with facts as they are today. Our policy should be one of permitting the employers' viewpoint to develop, holding back at all times our agreement, subject to securing from the employers in return for each modification of our restrictions a greater job and wage security.

In large part, the suggestion here of direct bargaining was necessary to convince the international council of the ALA that union prerogatives were not just being handed over to the employers. Robinson must have recognized that no swapping was going to be possible at the JLAC meetings, and that their function was quite different, to provide a forum in which issues could be discussed without committing the participants to a fixed position. The points of significance in the Robinson statement are not related to a narrow bargaining strategy. They are: (1) the union was accepting a goal — "a stable, prosperous industry" — that had little connection with the working class or the trade union movement in general; (2) it was contemplating the removal of its direct (constitutional) controls over working conditions in order to reach the goal; (3) it was consciously building an employers' group with which it could negotiate.

The JLAC was broadened in the fall of 1944 to include two representatives from the NAPL. No other additions were made, in large part because there were at that time no other important national associations. PIA was only organized the following year, and its predecessor, the typothetae united, had not paid much attention to its lithographer membership.

The union hoped that the JLAC would in time bring a return of national bargaining. This was an idle dream; the developments which came to fruition in the postwar period eliminated even the Joint Lithographic Advisory Council.

Chapter **VIII** ✍

The System in the Postwar Period

There was a sharp break with the past during the postwar period. This was true in the jurisdictional dispute, in relations between the ALA and the employers, and in the results of bargaining. An underlying cause for each of these three changes was the rapid expansion of the market for lithographed products. In no place was its effect more pronounced than in the jurisdictional conflict between the IPP&AU and the ALA, which could well have become explosive but for the existence of the National Labor Relations Board. The NLRB had been in existence for over a decade, but it did not become a key actor in the lithographic system until market expansion made it so.

Role of the National Labor Relations Board

The initial reaction of the ALA to the unit determination powers given to the NLRB by the Wagner Act in 1935 had been to broaden the union jurisdiction to include all workers in a lithographic plant. In the late thirties the orientation to industrial unionism became so strong that not only unskilled workers in purely lithographic plants, but skilled letterpress craftsmen were sometimes accepted into the union. The ALA did not seek to organize these workers, but if there was no contest for the unit and the management preferred to have all workers in one union, then the ALA would take in photo-engravers, compositors, relief pressmen, bindery workers, or any other craft that was employed in the plants for which it had bargaining rights. In 1946 President Riehl referred to this policy openly in the convention:

Again on organization, I want to repeat what I said last night in the heat of debate, that this organization cannot go on record as going out to organize printing pressmen or any other definite branch, but we are to continue to organize, as we have in the past, whenever the opportunity presents itself to organize an entire plant that has these various branches in it which are not organized.[1]

This approach made sense so long as the lithographic process was primarily confined to the lithographic industry. But it was becoming apparent even before the war that lithography was going to expand rapidly in commercial printing plants, in book and periodical publishing firms, and in private plants, all of them places in which lithographers would be outnumbered by other classifications.

In a series of cases beginning in 1943, the ALA defended its right to continue representing lithographers in combination plants rather than having them made a part of a larger unit including letterpressmen. It gradually developed a concept of the "traditional lithographic unit," which the board found acceptable, and the ALA found quite easy to defend. The challenge came in 1945, when two cases came before the board which both the printing unions and the ALA were determined to win. Extensive testimony was taken, and the decisions of the NLRB in these cases became of central importance as precedents for the board and in determining the ALA approach to organizing.

Foote and Davies and Pacific Press. The first of these cases was the *Foote and Davies* case, in Atlanta, Georgia.[2] The ALA had represented offset pressmen and lithographic platemakers in Foote and Davies, a large combination plant, from 1939 until 1945. At that time the IPP&AU was successful in convincing the management that jurisdiction over these workers should be given to it. After this, the IPEU laid claim to the platemakers. The offset pressmen were fired upon refusing to join the IPP&AU, and the ALA filed an unfair labor practice charge against the company. In holding in favor of the ALA, the board made the following points:

(1) The jurisdictional award of the AFL was meaningless and without effect. It had not been enforced for thirty years, had not been useful in resolving disputes between the unions, and was in any case irrelevant in an unfair labor practices case. It was no defense for the employer to say he acted in accordance with the AFL jurisdictional award.[3]

(2) The employees in the lithographic department constituted an appropriate unit for collective bargaining:

> The offset pressmen and the platemakers do related work, are segregated from other employees, and are separately supervised. In addition to these considerations, our unit finding is premised upon the important factors emphasized by the trial examiner — the absence of interchange between offset and other pressmen preceding the charged unfair labor practices; the distinct skills and duties of the offset pressmen and platemakers; the history of organization and collective bargaining by the Amalgamated throughout the lithographic industry; and, more particularly, the 6-year history of collective bargaining between the

Amalgamated and the respondent on the basis of a single unit of lithographic workers.[4]

(3) The contract including the lithographic workforce with the letter-pressmen was for an inappropriate unit and therefore invalid.

The third point follows legally from the first two; these, however, were matters of fact, and in the extensive record compiled the ALA was able to establish a foundation in fact that has served it well for over a decade.

The key points which the ALA was able to prove were that the AFL jurisdictional award was irrelevant, that lithographic workers *traditionally* bargained as a unit, and that the ALA always had been and was still the only graphic arts union which represented them in any measurable degree. Two pages of statistics in the trial examiner's report on the extent of ALA organization in the industry were concluded with the following paragraph:

While the record does not disclose the number of offset pressmen affiliated with the Pressmen, it would appear from the record that the latter have organized as members a comparatively small percentage of offset pressmen.

These three findings were important because they would hold true for all future cases. The other findings, separate supervision, past history of bargaining in the plant and so forth, would change with each case, and in the *Pacific Press* case it was illustrated that these latter circumstances were of substantial importance.

In the Pacific Press, a large Los Angeles combination shop, there had been no prior history of bargaining with any union. The IPP&AU petitioned for a unit containing only pressmen, the IPEU for a unit of all preparatory workers, and the ALA for a unit of offset pressmen and lithographic preparatory only. The board rejected the ALA petition, citing four reasons for so doing: there is (1) no prior history of bargaining based on a separate lithographic unit in the company, (2) "no affirmative evidence that the pattern of organization and collective bargaining generally present in the industry has included the particular local [Los Angeles] in which the workers concerned are employed," (3) no evidence that the lithographic employees were "markedly segregated or centralized as were those in the *Foote and Davies* case," (4) evidence of substantial interchange between employees on offset and other presses. Quite clearly the board was still impressed with the evidence submitted by the ALA and lithographic employer experts who had assisted the ALA in both cases. The *Pacific Press* situation was described as "unique in the industry . . . The facts, unlike those in the *Foote and Davies* case, tend to emphasize the relation between offset pressmen and letterpressmen and the connection between platemakers and photo-engravers, rather than the tie between offset pressmen and plate-makers."

The ALA had had the employers helping them, but the letterpress unions had been supported in their presentation by counsel from the AFL. This was too much for the leadership of the amalgamated, and it disaffiliated from the AFL. In early 1946 the amalgamated joined the CIO, taking great care to spell out its jurisdictional position in detail, and was a member of the CIO at the time of the merger of the two national federations.

Impact of NLRB decisions on union organizing strategy. In a sense, the *Pacific Press* case tended to emphasize how sweeping was the victory in the *Foote and Davies* case. But it also drew sharp attention to the importance of keeping the ALA a strictly lithographic union if full advantage of the board's power was to be taken. This was immediately clear to some in the union, notably Benjamin Robinson, and in his address to the 1946 convention he took up the question:

Shall we organize letterpress workers? The answer is squarely "no," because our right to a bargaining unit of lithographic workers exclusively would quickly break down if we organized all types of pressmen and platemakers. Our traditional and present, and correct, form of organizing would be destroyed. We would lose the autonomy we have so carefully established, and lose bargaining rights in a large number of establishments.[5]

General Counsel Robinson enjoyed a unique position in the ALA, which permitted him to discuss policy with the leadership of the union and directly with the membership in a way that would bring sudden death to staff men in most unions. But his reasoning did not always prevail, and for a period the ALA occasionally granted membership to letterpress workers. In 1955, in open convention, Robinson attacked this practice vigorously, arguing that the ALA had to organize and bargain for lithographic workers only, and specifically had to avoid representing letterpress or other graphic arts workers. The NLRB, said Robinson, had repeatedly pointed to the traditional bargaining unit for lithographers in explaining its unit decisions. But the carve-out decisions depended upon the purity of *all* ALA bargaining units. "That conclusion," said Robinson, "is not simply a whim of mine . . . That is the key to maintaining our position."[6]

This tightening of organizing standards so that there would be no direct efforts by the ALA at any time to organize letterpress workers was paralleled by restrictions on the organizing of general workers. Yet in spite of this some recent cases have seen the opponents of the ALA, either employers or other unions, drawing direct and potentially dangerous attention to the fact that the ALA is not a craft union, and has numerous contracts covering nonlithographic workers. In the *Sutherland Paper Company* case, counsel for the company put into the record evidence that the ALA represented shipping clerks, packers, inspectors, bindery workers, janitors and so forth, in an attempt to show that the ALA had no right to continue receiving favored treatment from the board by having "traditional lithographic units" carved out for it.[7] A portion of the conclusions of the employer brief point up the issues:

In complete derogation of its own criteria for craft severance, the Board has permitted the Amalgamated to sever groups of employees merely on the theory that employees were engaged in a "traditional operation" or "process."

We submit that neither the facts, nor the law, nor the Board's prior practices justify the creation of a special exemption for the Amalgamated from the requirements of the *American Potash* decision. On the contrary, we believe that it is abundantly clear that sound public policy and the purposes of the act require the Board to minimize rather than encourage practices which seek to destroy established bargaining units and the stability of bargaining relations. This, the Board has recognized and sought to accomplish in the *American Potash* decision. The application of that decision to the instant case, we submit, would require the dismissal of the petition herein.

The ALA has been able to establish a unique position before the board, but the position is endangered when substantial evidence is presented that the traditional lithographic unit is only one of several under which the ALA has found it convenient to bargain. In the last several years, the majority of representation cases brought before the board has comprised situations in which the ALA is seeking to carve out a unit from an existing bargaining relationship, rather than protecting an existing unit from being lumped in with other workers. The latter case predominated until the late 1940's, and permitted the ALA to establish firmly the principle of the "traditional lithographic unit." Although *American Potash* modified the standards for a carve-out situation in 1954,[8] the standards remain more strict than those applied to the ALA, and the argument is persuasive that they should be applied to the ALA when that union is attempting to carve out a new rather than to defend an existing unit. *Foote and Davies* has served the ALA well in the nearly two hundred unit-determination cases that have been settled by the NLRB, nearly all in the ALA's favor. Nevertheless, there is a continuing concern on the part of the union to match its organizing efforts with its jurisdictional requirements, so that the legal position before the board is not endangered.

Organizing success of the ALA. Winning the right to an election held for the lithographic unit alone is only the first step in the process of representing the employee in question; the election must still be won. When the problem was protecting rights already established, this difficulty was minor. So long as the right to bargain could be secured, the matter of showing the board that lithographers wanted the ALA to represent them was little more than a formality.

In recent years it has become much more; it has, in fact, become one of the most important problems facing the union. As early as 1949 reference was made to the need for following up board victories with more consistent election victories.[9] In 1955 the international president reported on the elections won and lost, with about one third of the elections resulting in defeats.[10] In the following convention such a report was not made to the delegates, but in his speech Benjamin Robinson noted:

to our amazement, we lose election after election, even in shops where a majority of the workers are fullfledged members of the Amalgamated . . . What has just been said has happened in Connecticut, in Florida, in Wisconsin, in California, in Ohio, in Indiana — all over the lot. What is the answer? Whatever it is, you simply must find it.[11]

The problem was highlighted shortly after the close of the convention. Robinson had pointed to an immediate challenge during his convention speech:

[The management of the McCall plant and the Pressmen's Union] fought us very bitterly, and therefore, I am extremely happy to be able to report that on September 11 I received from the National Labor Relations Board in Washington their decision which unanimously gave us a complete legal victory and ordered an election of lithographic workers [applause].

Now the job is yours. Now your job is to win that election; and if ever — if ever the brains and initiative and power of this organization should be used to win an election, this is it.

The convention "arose and applauded loud and long" after Robinson's speech, but the ALA lost the election. This was not a small shop, nor were the others in which the ALA won the right to carve out but lost the election. In 1958 when the ALA left the AFL-CIO in a dispute over whether it should be permitted to ignore the no-raiding provisions of the AFL-CIO constitution, George Meany documented his charge of "very stupid leadership" by pointing out that in 1956 the ALA had picked up nineteen and lost forty-two members in its attempts to carve out units from existing bargaining relationships.[12] In fairness to the ALA, Meany had to choose his "typical" year with some care. Currently, the ALA is having more success in winning carve-out elections than in 1956, though its record is far from perfect.

The 1946 Negotiations and the Canadian strike. The ALA had altered its jurisdictional and organizing strategy in the immediate postwar period; it also changed its approach to bargaining. There was quite naturally concern that the postwar period would be a repetition of that of the First World War, except that there would be a great number of army-trained lithographers who would destroy the union's opportunity to protect wage and other wartime gains. The Chicago local conducted an extensive study to prove that there would be no labor shortage in the area, and even if the industry received back from the army no more than were called into service, there would still be a surplus of men.[13] The Joint Lithographic Advisory Council organized an extensive program for attracting the skilled men that had been lost to the services, but at the same time emphasized that few new job openings would be available. In the first postwar convention in 1946, New York's Local 1 brought in a fiery resolution condemning a wartime relaxation of apprentice ratios which had been permitted in the artist classification as a "heretofore unheard-of extraordinary procedure" that should be halted at once. It demanded that apprentices made under the relaxed provision be carefully screened.[14] The shorter hours drive that began immediately following the war was a part of this carryover fear of unemployment that had been with the union since 1930 and before.

The break with the past was sharply defined in the 1946 negotiations in New York City. After thirteen years of good fellowship and cooperation between the ALA and the LNA, many of whose members were in New York, Local 1 initiated negotiations in early 1946 on the basis of nineteen points, among them the reduction of the workweek from forty hours to thirty-five, an increase in take-home pay, holidays and vacation time, and added overtime penalty payments. All these demands were aimed basically at restricting the supply of labor at a time when there was no unemployment, when in fact there was still a shortage of labor in some of the key classifications. The men who negotiated the New York contract for the union were Blackburn, Casino, Swayduck, Hansen, and Stone. None had been active in the Joint Lithographic Advisory Council,[15] and none had been known in the union nationally even in 1939.

The international officers opposed the local in setting such stiff demands, and asked it not to take a strike vote on the question. President Riehl was powerless to change the local's position, and Vice-President Bruck was brought in from Chicago to counsel moderation, again without success. Robinson also opposed the local position, but in such a circumstance he was powerless to do anything. Local 1, dominated even then by Ed Swayduck, although the local president was Blackburn, was determined to drive for big gains. In this it was completely successful, but at the cost of bitterness between employers and the union which lasted for some time. A letter from Local President Blackburn to Vice-President Bruck summarized the situation:

Some of the members are pressing me to go all the way with their demands. Bill Riehl has shown me telegrams from the Council requesting that I avoid taking the vote for a strike.

I believe it would be safer to stop the rough stuff at this time, but I must be guided by those who are bending every effort to get the best of conditions for the organization, even though we might injure the feelings of some of the employers who have been our friends.

I know that several remarks may come out of this which will be used to turn some people against the Union; but, it is my opinion that if a fellow like Floyd Maxwell [secretary of the LNA] wants to sit in at contract negotiations he most certainly should have expected to receive some of the abuse that goes with collective bargaining.[16]

A settlement was reached that included a 36¼-hour work week, and was accepted by the local membership on April 9, 1946. With the hours broken in New York the shorter work week spread rapidly to other centers in the United States. Only in the south was there widespread resistance to the reduced hours. The strongest resistance to hours reduction was in Canada, rather than in the United States.

In late 1948 the ALA locals in Canada presented demands which after extensive negotiations were reduced to the following: hours reduction from 40 to 37½ with no loss in pay, an employer-financed pension program, a 10 per cent increase in take-home pay, and employers to take over full financing of a health and welfare program which many plants already had put on a fifty-fifty basis. The employers' best offer was a 5 per cent wage increase and extension of the health and welfare plan on a fifty-fifty payment basis to those plants which did not already have it. Conciliation procedures carried negotiations to June 1949, when the union was free to act, and immediately struck four plants in Toronto. Other Ontario employers in the Canadian Lithographers Association closed down immediately, and by mid-July some fifty lithographic firms employing eight hundred ALA members in Toronto, London, Montreal, Hamilton, and Ottawa were out.

The strike lasted for twenty-two weeks, and ended with the work week unchanged, a 17½-cent wage increase, and a somewhat improved but still contributory health and welfare program. The wage increase was about 10 per cent, and if it is considered that the settlement came a year after the expiration of the contract, with only 7½ cents retroactive, it was not a great victory for the union. Nevertheless, it was the first national dispute in its history that had not been clearly lost. In the Canadian strike all locals held firm, and the employers were forced to raise their initial offer substantially. The employers gained time since the issue of the shorter work week was dropped from the succeeding negotiations. In 1953 the eastern Canadian locals did negotiate a pension plan, and in the following year, with negotiations pushed to the deadline and strike sanction received from the international, the employers agreed to the 37½-hour work week.

By no means was the Canadian strike the only dispute with employers during this period; others occurred in St. Louis, Louisville, Boston, Philadelphia, and elsewhere. The ALA was driving for shorter hours, pensions, higher rates, strong overtime provisions, and protective clauses, in the event of challenge by other unions. In some cases a strike was necessary to determine whether the concessions would be granted or not.

Andrew Kennedy's philosophy of working out mutually agreeable arrangements with the employers, depending on cooperation and employer recognition of the advantages of a union devoted to building a prosperous industry, was completely discarded by the ALA, beginning with the New York negotiations in 1946. It was replaced by the policy of improving conditions, to use Ed Swayduck's phrase, "by the sheer weight of our strength and sheer weight as a good strong Union, by showing our strength and throwing it around." [17] Yet in other respects Swayduck was one of the few genuine heirs of Kennedy in the union. His quarrel with the international was initially with its failure to organize, later its failure to provide leadership, and always with its failure to press for uniform conditions. But his method of conveying these strongly held convictions was reminiscent of Philip Bock, and he came to be known as "the wild man from New York," admired as a fighter but not wanted as an international leader.

ALA Leaves the AFL-CIO

During the 1930's the ALA would probably have been glad to sign a truce with the printing pressmen on the basis of no raiding of existing membership. Even as late as 1946, the amalgamated was not interested in the odd lithographer organized by another union. By 1952, however, the situation had changed. When the question arose of signing the CIO no-raiding pact presented in the CIO convention of that year the ALA decided it would incur a net loss by so doing:

> It is necessary for us to look at the future in the light of what has happened in the past. The members of the International Council know that the various attempts to solve our jurisdictional disputes with the Steel Workers' Union, with the Paper Workers' Union and with the Rubber Workers' Union have been substantially unsuccessful.
> We are unable to know in advance into what other industry lithography will penetrate . . . What [the no-raiding pact] comes down to is that we are required to place into the hands of somebody other than the Amalgamated the decision of where we are to organize, how we are to organize, and if we are to organize . . . This right is basic to our life.[18]

When the AFL and CIO merged in 1955, the no-raiding principle which had been in effect since 1954 was written into the constitution of the new federation, and later a no-raiding agreement devised, which the ALA and a handful of other unions refused to sign. The IPP&AU signed and pointed an accusing finger at the ALA as one of the disrupters of harmony in the graphic arts. The pressmen were on fairly safe ground. For a number of years the IPP&AU had written its contracts with a clause covering offset work if and when it was to be put into a plant under its control. Such a clause was the basis of its claim in the *Foote and Davies* case. There is nothing wrong with such a policy; no union covers only the workers employed in a bargaining unit when the contract is signed, and the IPP&AU has never agreed that lithographic workers should be treated separately from letterpress workers. But because so much of the expansion of lithography is not in lithographic firms, the ALA cannot accept this reasoning. In a speech to the 1957 convention, Benjamin Robinson outlined in detail to the delegates the position the ALA leadership had taken and the reasons for it:

> Now, when all of that is said and done, why is it that the jurisdictional position of the Amalgamated is so different from most unions?
> There are about eight-hundred offset presses coming into our industry every year, many of them going into letterpress plants, box plants, toy factories, and these other industries.

Now we are not too concerned about that rare press which escapes our control, and no matter how wide is our organization, there will be these individual exceptions.

But, I don't know how you can ignore this flood of competitive equipment, and to make matters worse, that equipment in large measure is going into plants working forty hours a week with scales 25, 35 and 40 per cent below Amalgamated standards. And of course, without the fringe benefits which we have.

But the Amalgamated's problem is even more urgent, because in fact lithographic equipment goes now into plants which are already under contracts with the Printing Pressmen, the Paper Workers, the Toy Workers, the Steel Workers, or various other unions. Therefore you are faced with the job of not only organizing an unorganized plant; you are faced with the job of organizing such workers away from a union which has organized the plant, even if it never organized these lithographic workers.

That brings us head-on in conflict with one provision of the Constitution of the American Federation of Labor-C.I.O., which says that no union shall organize workers in a plant which is under contract to another union.

Now, again, and you can't ask yourselves this too often, why do we insist on organizing these lithographers who bring us into conflict with these other unions and the AFL-CIO?

And when you do, you have to come up with the same answer. We have to if we are going to hold our jobs. Now, look at the Paper Workers Union. We know as a fact, that the large 76-inch offset presses have been installed and are moving now into paper mills and box plants, such as Fibreboard, Marathon, Sutherland, Container Corporation on the West Coast and East Coast, and others. If you examine the rates and the conditions of work, you find that their lithography is being produced at rates 40 per cent below the rates which are paid by our organized shops.

A very simple conclusion — if those plants can continue to operate their equipment at rates so far below ours, you have an absolute guarantee that our existing contracts, folding box shops particularly, such as Stecher-Traung, Rossotti, Rochester Folding Box, U.S.P.&L. and others, will surely lose their work to these paper mills and box plants and with them go your jobs.[19]

Immediately after the Robinson speech a resolution was presented requesting authority for the international council to disaffiliate at any time. This was amended from the floor to require a referendum vote, and was passed unanimously. In August of the following year the ALA was called before the AFL-CIO to answer for a series of raiding charges brought against it and sustained by the impartial umpire. It chose instead to disaffiliate.[20] At the meeting on August 22 with the AFL-CIO the amalgamated had intended to propose an amendment for consideration of the federation's executive council that would permit it to continue to organize lithographers in any industry. In the actual presentation of the case President Canary apparently forgot his lines, and all that happened was an acceptance by Meany of an ALA letter showing a willingness to disaffiliate as a statement of disaffiliation.[21] The die had been cast long before the August meeting, but all the councillors were present at the meeting, and Canary's fumble added measurably to the friction between him and the council.

The reasons for acute jurisdictional disputes are also reasons for a working agreement between the contesting unions. Ex-President Canary had always tended toward greater optimism on this score than members of the council. In his resignation speech three months later he expressed regret that he could not have furthered this goal, but at no time did he openly challenge the jurisdictional position of the council. A point which has not as yet been discussed openly in the leadership of the ALA is the uncomfortable similarity between the current ALA position and that of the IPP&AU during the 1920's and 1930's. The ALA is attempting to take over present membership of other unions, so also was the IPP&AU. The ALA is doing it because it fears that competition in the product market from firms whose labor conditions are not controlled by the ALA will affect the working conditions of present members of the ALA; this clearly was the rationale of the IPP&AU. The ALA insists on following the work regardless of the industry in which it is done; the IPP&AU insisted on following the work regardless of the process used to produce it. The ALA is spending heavily on legal counsel and using a substantial portion of its time and energy in processing these legal issues — money, time, and energy which could be spent on organizing. The ALA's criticism of the IPP&AU was that it spent all its time harassing the ALA instead of organizing unorganized letterpressmen. And the ALA is by no means winning all its elections, though its record is not so bad as the pressmen's in the earlier period.

A solution to this, and to many other day-to-day problems of the ALA and other printing unions, is for them to combine into a single union. In the following section the past attempts at amalgamation will be reviewed, together with current efforts in this direction.

Amalgamation Proposals

"One big union" in the graphic arts has always been a dream to many in the printing unions, quite possibly to a majority of the union membership or more realistically to the leaders of a majority of the membership.[22] The ITU has historically been in favor of the re-amalgamation of the five printing unions; the IPP&AU has similarly desired a getting-together, with the important difference that it is federation rather than amalgamation that has been the goal. With the ALA, of course, it favored complete amalgamation.

The lithographic unions were successful in merging into an amalgamated body in 1915, at a time when the jurisdictional dispute with the IPP&AU was still unresolved. Samuel Gompers was in favor of the amalgamation of the lithographic unions, but he was also in favor of the Lithographers International Protective and Beneficial Association joining the International Allied Printing Trades Association, and of arranging some form of amalgamation which would dispose of the dispute of the LIP&BA with the IPP&AU and the IPEU. The recommendation of the adjustment committee required the newly amalgamated lithographers to split themselves between the IPP&AU and the IPEU, with a weak and unstable group of flat-bed and direct-rotary pressmen left as the approved membership of the ALA. The ALA refused to accept either the proposed plan of amalgamation or the jurisdictional award, but under pressure to offer something constructive, made the following proposal in 1918:

[The LIP&BA] hereby offer the proposition to the International Printing Pressmen's Union and the Photo-Engravers' International Union, parties to the controversy, to amalgamate with the two above-named Unions, or with either the one or the other of the two, providing the plan meets the approval of one of the unions and not of the other, under one board of international officers and under one set of laws.[23]

This was unacceptable to the other unions, but any effort to take direct action to enforce the plan of amalgamation was unlikely to succeed, and in any event was far less important than the struggle to secure the forty-four hour week which was then breaking out into the open. During the 1920's, discussion between the unions continued, the printing unions urging acceptance of the AFL award, and the ALA requesting admission into the Allied Printing Trades Association on the improbable theory that all difficulties could be worked out there.

In the late 1930's there were again efforts to find a compromise that would eliminate the jurisdictional controversy by forming some sort of connection among the three organizations — the lithographers', photo-engravers', and pressmen's unions. Little progress was made until 1941, when President Berry of the pressmen unexpectedly joined President Volz of the photo-engravers in agreeing substantially to the plan proposed by the ALA in 1918. Yet in spite of this a solution was not possible. The ALA refused to join with the IPP&AU under any reasonable conditions, and the IPEU was no longer sure it wanted the lithographers. As Elizabeth Baker has pointed out, "Had the Lithographers' 1941 plan carried the ALA into the IPEU as a unit, its members would have outnumbered the membership of that very closely knit organization by about three to one, giving rise to the question of who was absorbing whom."[24] Nothing but conference reports came of the attempt, and later the IPP&AU reopened its efforts to have the ALA join with it. During the war and postwar period the ALA also received a number of offers from the ITU to affiliate with it, and on one occasion so much attention by the membership was given to such offers and inquiries concerning this possible affiliation, that it represented something of a hindrance to ALA organizing efforts, and had to be counteracted with official statements that it was still the policy of the international to organize all offset workers, and would continue to be for the foreseeable future.[25]

Just as IPP&AU interest in the membership of the lithographers' union dated from the introduction of the offset press, an innovation which it believed would cut into the demand for the craft skills of its members, the interest of the ITU in having the ALA join it originated when it appeared that photo-typesetting might bring the two unions into direct conflict. Apparently the first open comment that the ALA might be invited to ally itself with the ITU came in 1938.[26] These offers became more concrete as jurisdictional frictions increased between the ALA and the ITU, a trend which became marked in the early 1950's.[27] Unlike the pressmen, for whom the ALA has always had a basic dislike that goes well beyond the problems of the hour, the ITU is respected both for its aggressiveness and its form of government. Personalities interfered with any close relationship prior to 1957, but in 1958, in the midst of a rather bitter (and continuing) fight between the two unions in the western Canadian provinces, a memorandum was signed which spoke approvingly of active cooperation, and the desirability of joint organizing and negotiating "when feasible." A joint fund of $100,000 was established, presumably to be administered by a joint coordinating committee, also established by the memorandum. The agreement is at least as remarkable for what it does not say as for what it does; one wonders, for example, in just what sort of case $100,000 is intended to be spent. Benjamin Robinson has described the situation as one in which the two unions were unable to agree at the outset on what sort of cooperation would be feasible, yet each was anxious to demonstrate that it meant business.[28]

President Swayduck of Local 1, in many respects the prime mover behind the attempt to establish closer relations with the ITU, has gone much further. In an informal talk to the Metropolitan Lithographers Association, at which Robinson was also present, he chided the general counsel for being so cautious in his references to the working agreement with the ITU, and said that he was *sure* that the ALA and the ITU would work something out: "we'll keep working 'til we do — yes, merger, if necessary." [29] Difficult problems need to be overcome before such an arrangement can be made, but the pressure of technological change is such that the present division of printing unions is no longer stable, and a merger of the ITU and the ALA would offer substantial advantages to both. The ALA would be assured that its employers would have a secure source of type, a matter of crucial importance as lithography continues to expand into publications printing, and the ITU could dispense with its expensive and not wholly successful program for retraining its members in stripping operations.

Recently employers have taken official notice of the jurisdictional problem, and made their dislike of the present situation clear. President Mendel Segal of the Printing Industry of America's union employers' section invited all printing trade union presidents to address PIA-UES convention in the fall of 1960, and then in turn directed his keynoting speech to the assembled presidents:

The welfare of our industry and the job security of all of us does not permit the luxury of uneconomical solutions to jurisdictional problems. *You must find a way to take employers out of the middle.* The so-called "cherished lines of jurisdiction" must be compromised. You should recognize that work processes must be resolved and performed within the limits dictated by what is most economically feasible — by whomever demonstrates the aptitude to learn quickly or perform right now. This means that certain processes should be performed in varying ways under varying Union Contracts — both as between cities, and even among different plants within a city. Admittedly, this is not as desirable as a single contract with a single union for all areas and for all processes, which is a Utopia, which, I doubt, can ever be achieved (unless all crafts are merged). And, since there seems to be little chance of the problem being resolved between unions, the logical answer should be based on pure economics. I am convinced that unless a peaceful and economic solution is found soon, we are all going to be seriously affected.

Resolving this problem is the great challenge to union labor, and it must be resolved before non-union plants get such a head start that we can never catch up.[30]

Relations of the ALA with other unions vary considerably. Cooperation has been possible between the ALA and the bindery workers, steel workers, rubber workers and the pulp, sulphite, and paper mill workers in organizing drives, just as conflict has been almost inevitable when the ALA has attempted to carve out lithographic units from existing plant-wide units. The ALA's position is inherently unstable, for its ability to follow the process depends entirely on the continued support by the NLRB of separate lithographic bargaining units. Forty years ago it would have had the support of the AFL, which at that time chose to support the IPP&AU in its request to break up existing bargaining relationships of the lithographic union to preserve craft unity among pressmen. Today, when the ALA argues exclusively for its rights to lithographers, the AFL-CIO has shifted to a firm support of existing bargaining relationships, and so the ALA remains out of step with the rest of the labor movement.

Postwar Leadership in the ALA

Andrew Kennedy's untimely death in 1939 placed the burden of responsibility on Vice-President Robert Bruck, which he carried for a short period until a new international president could be elected. The man chosen was William Riehl, who did not choose to be a strong leader, and in any event would have found it difficult with the change in the local leadership and the radical change in the council. In 1947 he declined to run and was replaced by John Blackburn, president of New York Local 1. In the following election in 1949 and again in 1951 Blackburn was opposed by the new Local 1 President, Ed Swayduck, who carried the New York local each time but was roundly defeated, nevertheless. Neither Riehl nor Blackburn were effective presidents and during their administrations the international council developed skill in running the union directly.

The years from 1949 to 1955 form one of the more interesting periods in ALA history; it is not at all clear why John Blackburn aroused such violent personal antagonism in many of the union officers who served during his administration. For some, notably Benjamin Robinson, it was probably the feeling that a second-rater was sitting in Andrew Kennedy's chair. For others, it was because he was taking credit (very graciously, for this was one of his vote-getting qualities) for the impressive gains that were being made. Without doubt the amalgamated was poorly led during the period, though one can scarcely accept the logic of some current leaders who give Blackburn credit for none of the success but blame him for all the failures during his term of office. It is true that the ALA indulged in several strikes which were reminiscent of the late 1920's in the lack of investigation which preceded their approval and the mismanagement in their actual conduct. Certainly the outstanding one was the strike at the large Western Printing and Lithographing plant in Poughkeepsie. The strike had the initial support of less than half the Poughkeepsie membership, and to salvage anything from the mess the international was required to disclaim representation of over three quarters of its former membership in the shop.

The Poughkeepsie strike and several others, notably those at Buffalo and Baltimore, convinced others besides Swayduck that Blackburn had to go. Swayduck had failed twice in his effort to replace him, and, being reasonably certain he would again be beaten, pleaded with George Canary to run for the office. Canary was the able and popular president of Local 4 Chicago, and had achieved prominence in the international for being president of the first local to gain the thirty-five-hour week. He had refused once before to run for international president, and agreed in 1955 only because he recognized that the troubles on the eastern seaboard were due largely to ineffective international direction.

The estimate of Blackburn's popularity was accurate. George Canary had a four thousand vote plurality in the New York and Chicago locals, but won by only thirty-five hundred votes in the election. Such was the bitterness against Blackburn in New York, that immediately after his defeat for international president he was very nearly expelled on a technicality from Local 1, of which he had formerly been president, and was thereafter denied the right to hold even the most trivial local office. George Canary was the first elected president from the midwest since John W. Hamilton in 1906. None of the council members had served under a strong president; all were in the habit of making policy themselves, and looking to Local 1 or General Counsel Robinson for leadership and direction. It has already been noted that a majority of the members on the council in 1955 were elected from a regional constituency, and they owed their election to their popularity as presidents of large locals, not to union-wide prestige as artist representative, pressman representative, or such. President Canary quite rapidly found that his unquestioned pre-eminence in the local hierarchy did not carry over into the international, and that his right to refer or not to refer to Counsel Robinson, for example, had to be earned and was not his by virtue of his election as president. Other difficulties arose, most notably the perennial friction between Local 1 and the midwest.

There are some quite remarkable parallels between the internal problems of the union in 1904–1906 and those of the ALA in 1955–1958. The question of moving the international office west was being actively debated, the international president was working out of his home city in the midwest rather than New York, and the resulting breakdown in communications was creating serious frictions. President Canary was not granted the privilege of expelling Local 1, as his St. Louis predecessor had, but in the policy conference held in Cleveland in 1958 he did take the opportunity to advertise his scorn for Local 1 leadership and members of the international leadership in a long and unexpected speech.

This speech was greeted with a rising ovation, which indicated that though Canary had lost the respect of the international council, he had not lost the respect of the membership. As local after local arose to criticize the council, it became apparent that the international council was in trouble. The atmosphere of the convention hall was clearly reflected in the words a council member addressed to the delegates:

I think that better than being up here on this platform that I should be in the witness box . . . I say that as a jury that you have a prejudiced opinion — that you are standing in back of an ovation for a speaker this afternoon [Canary] by your lack of ovation for the speakers that have spoken since — by your complete lack of desire to support your International Council and its actions.[31]

The memory of the Cleveland policy conference is still vivid in the minds of those councillors who survived the 1960 elections. It helped dramatize the need for strong leadership at the top, but it also made clear that even strong leadership is unacceptable if it is not responsive to local sentiment. One must concede that the local delegates did a remarkable job on their leaders. They censured them for not doing enough, and for doing too much that was wrong. And they made both criticisms stick.

The events in Cleveland demanded a change in leadership. As President Canary himself pointed out, "You are going to have a new boy, you absolutely are because, after what you have heard here, how is it possible for me to stay on this job?" President Canary was replaced by the senior vice-president, Patrick Slater, a man who had been president of the San Francisco local during the 1922 strike. President Slater was clearly an interregnum president; he continued in office only until the end of 1959, when he retired.

He was replaced as president by the international councillor and Toronto Local President Kenneth Brown, the first Canadian elected president of a recognized international union in over fifty years. Robert Bruck had seniority, Riehl had prestige and dignity, Blackburn had a winning smile, Canary had the necessary Chicago support to combine with New York and unseat him, and Slater as the most senior vice-president had (like Bruck) the right to step into the vacated presidency. Kenneth Brown seems to have been tapped for reasons more closely related to probable success in the presidency, and it is quite possible that the long drought in able leadership has finally come to an end.

Summary

It is possible to identify a cycle in the relations between union and management in the lithographic industry. In the first stage, unions are weak or nonexistent, unrecognized by management, and they have little power to affect the determination of wages, hours, and working conditions. In the second stage, they gain in strength, and by conciliatory and responsible attitudes convince managements that there is more to be gained by dealing with them than by ignoring them. In the third stage, managements decide that too much is being demanded by the union, and attempt to reassert their rights by an organized attack on it.

This cycle has been repeated several times. The first ended with the long strike in 1906–1907, the second with the 1922 strike, the third in 1927, and the fourth in 1946.

From the organization of the union until the early 1890's both worker and employer organizations were weak and relatively unimportant. Then for a decade the national worker organizations exercised substantial power, and sufficiently misused it in the years 1902–1906 to impel key employers to form a strong national organization, initially to deal with the unions but later to destroy their power in the 1906–1907 strike. The second cycle began in 1907, and union weakness continued until amalgamation of the unions in 1915 and full employment in the years immediately following gave them sufficient power to introduce national contracts again to the industry.

This, as in the period 1904–1906, was genuine national bargaining. The international officers of the union and the board of directors of the association met together, negotiated, and came up with an agreement which was applied throughout the country in association shops, and became the pattern for all nonassociation shops. Nationally uniform wage levels were not negotiated, but nationally uniform wage changes were, and so too were procedures for settling other issues which would arise between the parties. During this second period of national bargaining the use of flat increases (not percentages) had the effect of ironing out regional differences, since the agreement was in force during the period of the rapid increase in wages during the war and immediate postwar period.

The end of the second cycle came in 1922, when the employer organization successfully resisted a reduction in hours and insisted on a reduction in wages, the resulting strike again decimating union ranks. In the third cycle there was no second stage of union-association harmony at the national level. The culmination of this cycle in the 1927 shorter-hours strikes was a much less decisive victory for the employer position than was true in either 1906 or 1922. Yet from the turn of the century to 1927 it is clear that labor relations problems were dealt with as national questions, to be determined by reference to national contracts, the union constitution, or the LNA *Shop Rules,* but determined nationally in any case.

The Great Depression and the NRA introduced a period in which the cycle is less easily identified. The union exchanged a trade union philosophy for an industry-oriented philosophy, and government entered the industrial relationship, first as a sort of scorekeeper and part-time umpire during the NRA, and later as a rule maker and full-time umpire with the Wagner Act. The pressure of unemployment and the opportunities presented by the NRA encouraged the printing unions to attempt to weaken the lithographic union, and these same opportunities endangered the power and prestige of the lithographic employers' association, and to some extent the welfare of the lithographic employers. A rather short first stage of union weakness (1929–1933) was followed by a long stage of labor-management cooperation from 1933 to 1946.

Other factors at work during this period fundamentally changed the nature of the union-management relationship. Technological changes caused lithography to enter new markets, rapidly increased the number of firms using the lithographic process, and reduced the size of the typical lithographic firm with attendant changes in the ability of lithographic employers to co-ordinate their activities and in the ability of a single employers' association to secure uniform labor costs throughout the industry. Proportionately, the greatest expansion of the process was in local markets, served by firms that did not affect and were not affected by labor costs in other geographic areas except through labor market pressures transmitted by the union. Initiative for maintaining labor cost uniformity passed to the union from the employer organizations, which had grown to three in number and had decreased in strength proportionately.

Although the NRA reintroduced a period of national determination of labor questions, these market and technological developments made national bargaining without government support impossible. The failure of the 1937 national contract showed conclusively that the period of national bargaining had ended in the lithographic industry. The question did not arise as to the representative character of the LNA, for it was unable to commit even its own members to a contract approved by the board of directors.

The same market and technological developments which brought about the end of national bargaining also led the ALA to switch from a defensive to a vigorously aggressive jurisdictional position. From 1913 to 1946 it was seriously attacked only by the IPP&AU. From 1946 onward, by continuing to adhere to its policy of organizing lithographers, it came into conflict with a series of unions, most of them industrially organized, with which it had had no contact before. The old slogans of industrial organization, useful during the 1930's, were quietly discarded, or used only insofar as they strengthened or sustained before the NLRB the right of the union to a "traditional lithographic unit." The NLRB, concerned with the rights of workers to organize themselves, granted the ALA the right to hold elections in this unit. The AFL, the CIO, and then the AFL-CIO, concerned with the rights of unions to hold what they had, opposed the ALA, and their position led eventually to the ALA's current independent status.

With its primary enemy now other unions, the ALA would like to have the cooperation of employers. In general, it has the applause of employers when it organizes a low-wage competitor, but since 1946 it has not received any cooperation. For the fourth cycle ended in 1946 when the union placed a series of demands on New York employers which the industry considered outrageous. But instead of the reassertion of employer dominance which occurred in 1906, 1922, and 1927, the New York employers, and soon those of all other sections of the country, conceded to the union demands. National employer unity was no longer possible, either for or against the union.

This, then, is the historical context of the current union-management relationship. Bargaining is local, but with strong pressures for national uniformity. The union is stronger at a national level than are employers, but there is no longer one union only which actively organizes lithographers.

Chapter **IX** ✆

The Machinery of Bargaining

Collective bargaining has moved to the local level in the lithographic industry, in spite of the best efforts of both employers and union to keep it national. In the immediate postwar period this development was still being lamented by employers, as the following exchange in the LNA convention illustrates:

Mr. Wolff: When the union comes to us, they tell us what their national office will permit. Why can't we tell them what our National Association will permit?

Mr. O'Brien: Frankly, because their national office has a discipline over its members that this Association doesn't have . . . You see, the trouble with industry bargaining, of course, always is that you make the bargain the weakest member will take.

Question: How could we strengthen the employer organization in our dealings with the union on collective bargaining?

Mr. O'Brien: At the risk of being a little brutal on that, I am going to try and answer it. I don't think you can do it anyway except improving the backbone of your members [applause].

Mr. Hirsch: Speaking of this area arrangement as against the individual plant negotiating with the union, in my opinion it is greatly preferred. We have it in San Francisco and we are glad we do have it. However, it is certainly not the answer, because as New York goes, so goes the nation; but by the same token, as San Francisco goes, so goes the nation or any other sizeable area. The domino teeters; if it falls the wrong way, down they all go. Does that not point to the wisdom of having a national contract rather than area or individual contracts? Because New York was put through the wringer just a short time ago and we are all going to get it.[1]

In 1946 there were still employer dreams of a national contract and a strong association. The LNA had retained a labor lawyer as general counsel, and an expert in industrial relations, George Mattson, was hired as director of a new "labor relations service."[2] The ALA was suitably impressed; George Mattson was described as "some crackerjack anti-union fellow" with stacks of money, busy from morning to night gathering statistics and feeding information to employers that would permit them to hit the union where it was weakest.[3] Unfortunately for the employers, the information and assistance did nothing to build a national resolve to resist the union. The most useful information could not induce an otherwise reluctant employer to take a strike for the sake of a competitor, even though he realized that if all employers would do so, he himself would be better off. The LNA effort to take a firm stand gradually dwindled, and the labor relations service became another fact-gathering and -distribution office, soon to be paralleled by competing services of the two other employer associations.

In time, realism prevailed, and a more common sentiment among employers today is to applaud the very lack of uniformity which was once the central purpose of the LNA. One LNA member has stated the reason with great frankness: "If the LNA were in on the bargaining, or did business at the international level, they could not talk out of one side of their face in one area and out of the other in another area."[4] It is apparent that as the bargaining power of lithographic employers has decreased relative to the union's, the best strategy has changed from a united front to atomistic bargaining, at least to the point where the cohesiveness of the employer unit is nearly equal to that of the workforce.

National employer coordination of bargaining strategy is now recognized as a two-edged sword; regional coordination less so. On the eastern seaboard there has been little interest in coordinating with New York, which, particularly in the early postwar period, led the nation in high wages, high fringes, and extensive union participation in issues which are still considered managerial prerogatives. The same situation applies to Chicago, though in a lesser degree, because Chicago is not so far ahead of the pack as is New York. There is a tendency toward close cooperation among the Toledo-Cleveland-Akron employers and the Cincinnati-Dayton-Columbus employers, and coordination of strategy between the two groups. In 1947 the union believed there was a Cincinnati-St. Louis-Kansas City axis that was determined to stop the shorter hours, and an agreement that any shop struck by the union would receive financial assistance from other firms.[5] This is probably inaccurate, but it does appear that close coordination of strategy was taking place.

Throughout the 1950's there were formal and informal attempts to establish common positions on key issues before major cities went into negotiations. In 1953, for example, the LNA made an especially strong effort to have lithographic employers stand fast against thirty-five hours. One employer recalls a telegram sent to a firm in Chicago, where the key negotiations were under way, "urging them to hold the line and offering a vaguely stated sort of assistance if they had 'trouble.' "[6] During the late 1950's there have been several semisecret conferences between representatives of key employer groups for the purpose of establishing a common position on the issues of the hour. In 1955, for example, the complement of men on presses was such an issue, and, in 1958, "language," the protective clauses devised by the ALA, was another vital one. These attempts at coordination have had some value, but fall far short of the results secured in the first three decades of the century, when a truly unified employers' association determined employer policy toward the union. Although other employer efforts to build a united front may have occurred, the results do not indicate that they have been very successful.

Union strategy. On the union side the picture is a good deal clearer. In 1946–1947 the rapid expansion in the demand for lithography gave the ALA an opportunity to secure substantial concessions, and the break-through in New York on hours, wages, and fringes in 1946 supplied other locals with both evidence that they could and international pressure suggesting that they should do the same. A concerted effort brought shorter hours, either 36¼ or 37½ to 80 per cent of the membership within one year after the initial negotiations in New York.[7] Achieving this widespread extension of a pattern set on the eastern seaboard so quickly was not accidental, as the following story of negotiations in four central region locals illustrates. The quotation is from a much longer talk by General Counsel Robinson at a meeting of locals going into negotiations. It was intended to inspire, and doubtless some of the references to organized employer resistance were exaggerated.

Now, I want to paint the picture right now of how we broke these locals in the Midwest and the employers, and I want to show you this picture because I think it shows more than anything else I can say, the inter-relationship of negotiations in this industry.

Chicago made its deal in January. Chicago got an over-all increase of between 20 and 21 per cent, as we figure it, George. George [Canary] was not present in Chicago and we weren't very happy about that deal; that is, the International. But we realized they made the deal and that was that.

Meanwhile, we had been pushing Boston. As a matter of fact, they had a meeting and [President] John Blackburn and [Local 1 President] Eddie Swayduck went up and spoke to Boston, and I might say that in Poughkeepsie we did the same thing. We had a couple of meetings up there and the same fellows were up there. [Central Region Vice-President] Ollie Mertz was up there one day and I was up there a number of times. We were pepping up the local membership before we started our fight and while we were fighting with the employers.

We were in Detroit and we were heading for negotiations in Cleveland. The Cleveland employers had pulled a stunt which some locals fall for. The

Cleveland employers had offered six dollars across-the-board to the Cleveland local, with no reduction in hours, no increase in vacation, no new overtime rates, and the negotiating committee in Cleveland was recommending it for acceptance to the Cleveland membership. We begged them by long-distance telephone not to go to the membership with that offer until International representatives could get into town.

We went to Detroit first, trying to pep up that local, which, unfortunately, is not one of the strongest locals, and while we were in Detroit, Ollie said, "You know, there's a meeting in Toledo."

Now, Toledo's lithographic industry is substantially Graphic Arts Plate-Making Company, the largest plate-making trade shop in the United States.

Word had been sent to all locals by wire — I am sure you remember it — not to use any plates or negatives or positives coming from the Graphic Arts. We had learned by that time that the employers in the United States had started to hold meetings in various central cities for the purpose of developing a united front against the activities of the Amalgamated, and they were determined to hold the fort in that Miami Valley, Middle West area. Ernie Jones, the owner of the Graphic Arts plant, is a prominent member of the Board of Directors of the Lithographers National Association.

We got to Toledo about eight o'clock. I remember it so well. We attended the local meeting and spoke there, and I think it was about ten o'clock in the evening when I said, "Let's go and see Ernie Jones. Let's get him out of bed and talk to him. What the hell, we're up all night; let's do the same to an employer."

Well, we walked out of that plant at half-past two in the morning with a very famous document, signed by E. E. Jones. That is the contract we drew in Toledo that night. We got in Toledo — and this was unheard of in that area — eight dollars for everyone getting $50 and up, six dollars under $50; 37½ hours immediately and 36¼ hours by July 1st; we got the new overtime rates, double-time all day Saturday and double-time after two hours, and that was very important in Toledo because they were doing a lot of overtime work; two weeks' vacation after one year; and six paid holidays.

Ernie Jones said, before we left, "I guess the employers in the country will crucify me," and we said to him, "If you can show the Amalgamated that employers are boycotting you as a trade shop because you gave us this deal, we'll take care of those employers." So about three o'clock in the morning, we got on the train at Toledo on our way to Cleveland. We met the local negotiating committee. We went into session with the Cleveland employers — and they're really tough in that town. Those are the toughest guys I've met yet, until I got to St. Louis . . . [The employer spokesman said,] "You are asking us to give you an increase of over 25 per cent. Well, let's get down and talk 10 per cent. That's what we can talk about, and no shorter hours." He said, "We're not interested in what happened in New York, we're not interested in what happened in Chicago, and God knows we're not interested in what happened in San Francisco. We're interested only in the State of Ohio."

I was kicking Ollie's shins to keep him from laughing, and I said to him,

"I think you're absolutely right. I don't see what the hell they're doing in New York has to do with Cleveland. You ought to follow the pattern in Ohio."

By that time, the employers were getting a little bit worried, so I asked Ollie to let me have the contract which we had just signed in Toledo. I said, "Well, I'm going to tell you what the Ohio pattern is — signed at two-thirty this morning," and I told him what it was, and there was utter silence. You know, I've got to get some fun out of this job.

You see, the reason why the employers were so stunned was that they had an agreement among themselves not to break the 40-hour week in that area and not to give over a 10 per cent increase on scale. Out in those areas, that is something new. They don't talk about an increase across-the-board. They talk about scale increases. That is why you don't find as many premium men out there.

Well, we made our deal in Cleveland. We went to the membership of the Cleveland local that night and they were kind of ready to take a deal without a substantial increase in pay, without a cut in hours. We told them about the Toledo deal and that just electrified the whole local. The local, to a man, stood up and threw back the employers' offer.

Then we went on to Detroit. We made a deal in Detroit. We were able to get our members to back us and we were able to break the employers in Detroit because of what we had done in Toledo and then in Cleveland.

Boston, meanwhile, had settled. They had broken the hours to 37, but we had decided we weren't going to sign any Midwestern contracts for less than 36¼. Bircher was working in Erie and I think we kind of persuaded him that he could get the same deal there, and we eventually got that with some minor variations.

Well, so much for those Midwestern locals. I am positive, as positive as I am that I am standing here, that you would never have made a deal in Erie if you hadn't made the deal in Cleveland, because Erie is a poster shop and you have two big competitors in Cleveland, Continental and Morgan. You would never have made the deal in Detroit if you hadn't made the deal in Cleveland, and I don't know what would have happened in Cleveland if we hadn't made that surprise deal in Toledo.

It is evident that the union was conducting a loosely knit national bargaining program, and a closely coordinated regional one. The employer organization was purely defensive, and remarkably ineffective. It is also evident that Robinson was doing the directing and much of the crucial negotiating. He was also leading the union meeting intended to continue the spread of the national pattern. The general counsel was unquestionably the leader of the union during this period.

A further point may be noted. Much of the time spent in negotiations by the international officers and staff was in "pepping up the local," "pushing Boston" and doing the same thing in Poughkeepsie, and so forth. Like the employer associations, the union has been faced with a problem of securing member support for a uniform policy, and, since the war, for an aggressive policy. One of the most touchy problems has been convincing locals that they were not being asked to fight for the sake of Local 1. The antipathy for New York has been described; it has not been difficult for employers to use it to add a cautionary spirit to a local. When Robinson arrived in Cleveland to address the membership, he was told not to "shout too much because the employers have spread the word around that you are a radical from New York, so be kind of quiet with the men." As Robinson noted, "We were very quiet. That propaganda . . . seems to be very effective in a lot of locals."

A second problem is convincing locals that they can win and that they ought to support vigorously all demands for major concessions. The extensive surveys conducted by the international, for example, are primarily designed to emphasize to the members the disparities that exist between locals, not to make a better case to present to the employers. When a local is in negotiations, the greater danger from the international viewpoint is that the leaders will present and the members accept a package that will make negotiations difficult in other parts of the country. Sometimes the problem is a local president not interested in coordinating negotiations, more often it is a membership not willing to support a stiff set of proposals down to the wire. When the local leadership is the problem, there is little the international can do. Locals have been known to eject international officers forcibly from meetings when their presence was not requested (the Atlantic region vice-president was once so treated in Philadelphia). No union can afford to have many such incidents. Sometimes even a local leadership that agrees with the international cannot do much with its members. Milwaukee in late 1946 illustrates this point; the local president had to fight to get the council board to approve the use of a wage-reopening clause. International Vice-President Mertz helped to convince the Milwaukee Council Board and the negotiating committee that it was not sinful to ask for more money:

I never will forget that Sunday afternoon . . . when I insisted that we had to have more money, they insisted that I get out of town, and I insisted that I wouldn't. It finally went to this point, that they went before the membership and took a vote as to whether they would allow me to come in there to open the contract for more money.

That is the long and the short of it and, believe me, that's a tough bunch of hombres to do business with.

The problem of coordinating negotiations continues to be difficult. In almost every convention there comes a time when some local accuses another of "letting the ball bounce," of settling for conditions that are believed to be less than could have been secured short of a strike. The accusation usually accompanies the equally regular proposal to give the international more power in the negotiating process. The 1959 convention finally approved such a resolution (later ratified in a referendum ballot) which gives the international the right to review and approve proposals before they are submitted to employers, the right to address the membership before any negotiated package is submitted to them for approval, and which makes all contracts subject to international approval.[8] "Local autonomy" came in for an awful drubbing during the convention debate, particularly from that meek and docile follower of international policy, President Swayduck of New York. The apparent contradiction between Swayduck's well-known disagreements with the international and his opposition to "local autonomy," however, is superficial. His long-standing criticism of the international is that it lacks a well-planned and aggressive negotiating campaign throughout the country. Opposition to his definition of local autonomy has been quite consistent with such a position:

> We have locals, trade shop locals in our organization who have the audacity to speak of local autonomy when their rates are 30 and 40 per cent below some of the big locals that have been sweating and working and spending fortunes getting rates up there.
>
> How can they sleep nights knowing their wares are coming into these locals and throwing people out of work? That is scabbing. They have no right to scab under the banner of local autonomy.
>
> And, this is what we are a union for, the member, not his [the local president's] ego.
>
> This local autonomy means: "Me, I am the President. You can't bother me." The hell with you. You represent a member. That is what your a union for.

Among employers, few convention decisions attracted more attention than the one to approve the resolution on the "coordination of negotiations." Several thoughtful employers with long experience in dealing with the union now see it as a major step toward centralized decision-making in union bargaining tactics.[9] But such a conclusion goes too far. International approval has long been required for local contracts, and the only significant changes in this resolution are prior approval of proposals and direct access to the membership. Moreover, the 1959 convention also approved a resolution giving top priority to organizing, with explicit recognition of the fact that international representatives would have to spend less time servicing locals and more on organizing.[10] International people can hardly spend more time on both, and so the practical results of the much-debated "coordination" resolution remain in doubt. It is quite probable that persuasion will continue to play a more important role than compulsion. The process of passing the resolution was more important than any formal power it has given the international. This should not be misunderstood. There may well be greater uniformity among local bargaining positions in the future. But the uniformity will not come from a top-down pressure, but from an acceptance by the membership of the fact that such uniformity is in their interest.

The Temper of Negotiations

The temper of a negotiating session is in large part determined by the degree to which parties are separated on the same issues. If each has different issues which are most important to it, then a solution can be found in which both can be better off, in the sense that each loses a little but gives a lot to the other party. For example, recent years have seen a heavy emphasis by the union on "language," a generic term that for the most part deals with union security and rights to jobs. This emphasis often has had the effect of making locals, as one leader expressed it, "give a penny here and a penny there" in order to insure that they get the language.

Negotiations are stiffest when the hierarchy of issues is ordered in the same way by both union and management. At the time when Local 1 determined to create additional demand for labor by adding to the vacation period and reducing hours, New York employers would have been far more willing to raise wages than to reduce the straight-time hours available to them, and the bitter 1946 negotiations resulted. The same issue given high priority helped trigger the Canadian strike, although the fact that the strike continued after the hours demand was dropped indicates that there was basic conflict over the money and prerogatives issues as well. Chicago witnessed negotiations that are well remembered by both management and the union when the hours were reduced to thirty-five in that city during an economic period which both the union and management now agree would have kept the membership employed full time. By contrast, when the union and the local management group are willing to follow a pattern set in other cities, the issues remaining can usually be sorted out in harmonious negotiations. This sort of bargaining is typified by two passages from interview notes, the first from a union and the second from a management spokesman:

One time I went into negotiations and said, "These are 8 proposals we'll settle for. We've got 20 more we can bring in, but here's what we want." And in 6 weeks we had a contract signed. Some modification, of course, but we got all 8 established.

We have prenegotiations sessions. The union will bring up something the members have proposed from the floor, and if it's way out of line I'll tell him, "You better knock that down." If he can't he's got his work cut out for him.[11]

The union spokesman here represented a local not considered one of the strongest, and the management spokesman headed the bargaining group in a city in which the union is very strong. But in neither of the negotiations to which they referred was there a single issue which both parties placed foremost, and on which there was basic disagreement.

Bargaining is complicated by splits within the parties. In the Boston negotiations the employer group was put under severe strain when the union drove hard on the question of getting a four-man complement on four-color presses, which affected only one shop with seven such presses. Other shops, having only two such presses among them, already had the complement. Local 3 Boston offered to sell the complement for better economic conditions, which would affect all shops. The effect of this on the employer group was best summed up in a statement by Douglas Reilly, head of the employer bargaining committee, to a joint meeting of union shop delegates and management representatives:

I want to spread this on the record and I don't care who knows it that there was a point in the negotiations when it was pretty obvious that what we were looking for for about ten or twelve years was about to go in the ashcan and I think you know what I am talking about; it was the complement of men. I went to one of our other employers and said, "Listen, we don't feel we can go along on this situation any longer." They said, "We don't feel that we can either." Then it was a question of whether the management group was going to founder on the rocks or whether it was not. So I got up in our own management group and I said, "If you fellows can sell the idea of flexibility [not requiring full complement at all times] we [Reilly's company] will go along with it to that extent." In other words, we put a competitive disadvantage up on the block rather than see the management group and this collective bargaining effort we have had here go on the rocks.[12]

The writer was present during some of the employer conferences in which the issue was discussed, and there is no question that the unity of the group hung by a thread. It remained as a unit not only for the reasons the speaker gave, but also because the union compromised on its own position. Both concessions were necessary for the continuation of group negotiations.

In other cases it is the union which is troubled by internal disagreements.

Sometimes these splits have their humorous side, like the one in Boston which brought to an abrupt close a discussion of the same complement issue which split the employers. The union committee was made up of four men, one of them Bill Cowhig, a first pressman at Forbes, where the three-man complement was in effect. At the twenty-fifth negotiating session another member of the union committee waxed eloquent on the tremendous amount of work the fourth man had to do, detailing all of his various duties. The following is from notes of the bargaining session:

Doug Reilly: You know, it just strikes me now, if the fourth man is doing all this, what's the second pressman doing all this time?

Bill Cowhig [*interrupting Curtin*]: By —— that's what I want to know. I'll tell you one —— thing, he's not working for me, I'll tell you that. I don't give a —— but when Doug said that I couldn't hold it any longer. That's what I want to know, what the hell's the second pressman doing all this time? He wouldn't be working on my press, that's for —— sure.

Other differences raise more serious issues, because splits within the union have to be handled with great care. If an employer leaves the group, the pressures which brought him in in the first place may cause him to return. But when members split from the union, they may go to another union, never to return. Such a case occurred in St. Louis, when during a strike several plants went over to the IPP&AU, giving the ALA local in that city a problem which continues to plague it. It is one thing to follow the national pattern on rates and hours when local competition is with letterpress printers, quite another when numerous local lithographers are organized in a competing union.

Informal and Noncontract Negotiations

When issues are complex, when personality clashes are involved, or when there is a split in one of the parties which threatens the stability of the negotiating relationship, informal negotiations are sometimes used. In Boston during the last negotiations there came a point at which an impasse was reached because of the uncertainty of the employers concerning the approximate size and nature of the package for which the union would settle.[13] Such a situation called for an informal discussion between Al Heubach (employer) and Arthur Willis (union) to clarify internal disagreements which could not be discussed at the table. Much the same sort of discussion took place between Chicago President Canary of Local 4 and James Armitage of the employers' group in Chicago during the thirty-five-hour-week negotiations in 1953.[14] In a perceptive article on the process of bargaining, Meyer

Ryder has pointed out that although each side commonly has built a substantial "cushion" into its bargaining proposals, the tactics necessary to conceal this cushion may lead to an unnecessary strike:

> In the lower range of their respective bargaining cushions, [the parties] will buttress their simulated real positions by advancing the impression that they are at breaking-off points where further negotiation is fruitless. These are called false breaking-off positions and are utilized to compel agreement at some favorable point in the claiming party's cushion.
>
> Should negotiations actually crystallize into a strike at the false breaking-off point of either party, it is indeed unfortunate and indicates lack of skill in the party that has moved to the false breaking-off point. Skill is demanded in the manner in which negotiators create false breaking-off points to prevent their conversion into real breaks in negotiations. Avenues of retreat to the real bargaining position must be available, though not indicated, where a hoped-for victory is pressed strongly at a false breaking-off point.[15]

In lithographic bargaining it is a lack of unity within each bargaining team, as often as a lack of skill, that may lead to a real break in negotiations before all avenues of compromise have been explored. When contention involves an issue about which both sides feel strongly, it may be necessary for communication to be unhindered by an audience greater than one, if strikes based on misunderstandings are to be avoided.

In addition to informal discussions between the union and management spokesmen, other less conventional methods of reaching agreement are used, as when General Counsel Robinson and Vice-President Mertz closed the contract with the large Toledo employer at two in the morning, or when an international representative and a local president closed a deal for shorter hours in a northeastern city without the membership's knowing it was going to be requested.[16] Normally such negotiations occur because of the well-founded international conviction that locals will often settle for less than they could get. Yet such a practice is not without danger, particularly when the precedent is applied to the parallel case of the local's ignoring the international. In the same transcript that reports the two instances above, there is reference to a third negotiation which was completed over one weekend in spite of an international request to hold off a settlement. The local union president was promoted to foreman following the conclusion of negotiations. The international believed, probably unjustly, that the local president had sold the union down the river, but the incident did point up the danger of one-man negotiations.

Another type of negotiating away from the bargaining table is typified by experiences of the Joint Lithographic Advisory Committee during its brief existence. Since 1946 there has been little national discussion of issues with the intent of reaching firm conclusions, but three may be mentioned.

There has been a continuing contact with the Lithographic Technical Foundation to cooperate in the establishment of training programs. Both the foundation and the union recognize the importance of having effective training facilities if new methods available to lithographers are to be used quickly, and so there is room for a sort of arm's-length working arrangement between the two. More recently the ALA attempted to set up national apprentice standards which would be jointly supported by the union and the three associations (Lithographers National Association, National Association of Photo-Lithographers, and the Printing Industry of America). The IPP&AU has a national agreement with the PIA covering apprenticeship regulations, and the ALA believed a similar one would be beneficial in the lithographic industry. Apparently the LNA was only slightly interested, and backed off before discussion got well under way. The NAPL attended one meeting after which it, too, ceased to take an active part, and discussion was then only between the union employers' section of the PIA and the amalgamated. No agreement was possible, and the ALA settled for a unilateral filing of national apprenticeship standards with the Department of Labor's Bureau of Apprenticeship.

There was wisdom in this move. A governmental document which reproduces a private statement carries greater weight than the same statement from the original source. But the tactic may, as in the apprenticeship case, put an end even to informal bargaining. The final example of informal bargaining, revolving around the "one million dollar offer" illustrates this clearly.

Prior to the 1957 ALA convention a manning committee was set up in the union to determine appropriate complements of men on new equipment coming into the industry. In general this function is left to the employer, but the ALA, faced in particular with the problem of appropriate manning of web-fed equipment and with the fact that employers throughout the country were not consistent in their manning decisions, chose to do some investigating on its own. This committee became the committee on technological developments, and in its investigation of new equipment then available or soon to be so, recognized the remarkable impact the introduction of this machinery could have on the industry, both in expanding its market and radically altering labor force requirements. Local 1 President (and International Councillor) Swayduck conceived the idea of a great joint effort of the ALA and the employers to "harness new technological developments for the mutual benefit of the industry and the consuming public." [17] According to a *New York Times* interview, he took to the 1957 ALA convention the idea of a two-million-dollar fund to promote automation in the industry, the cost to be shared equally by union and management. According to the *New York Tribune* interview after the convention, the delegates were so enthusiastic that "they were all for voting $3,000,000." [18] This is, to say the least, a highly impressionistic version of what happened. In sum, Swayduck suggested the motion approving jointly operated funds for research, it was made and seconded, and he then suggested a sum of money, "let's say three million dollars." This was criticized as excessive; he pointed out that it was the principle of cooperation that the motion endorsed, "let's not go haywire about taxing ourselves." [19] The motion then passed without further discussion.

Six news items and four editorials later, the ALA initiated discussions with the employers, with predictable results. One million dollars was not readily available to any (or all) of the employer associations, and none was anxious to commit itself financially to an organization that would be partly controlled by the ALA. Moreover, the purpose of the fund was never explicitly stated by the union. It was clearly to do good technologically, but there was never a clear and precise statement of intent, as there had been with the apprenticeship program. Actual discussion, to the writer's knowledge, never went beyond informal questions. An interview with a trade association executive in January, 1958, expressed the prevailing employer attitude:

What are their motives? That's what we don't know . . . What if they offered $6, 10, 25,000,000? What's the difference? They know we haven't got that kind of money. It's just good publicity. The Lithographic Technical Foundation is an employer-financed body and we're going to keep it that way. They're already doing a good job.[20]

Neither the apprenticeship nor the technological developments programs have achieved the degree of success desired for them, or which might have been possible prior to 1946. A compromise for the apprentice program might have been worked out if the union had been willing to devote its full attention to the negotiations, but the million dollar plan had two strikes against it from the start because of Ed Swayduck's penchant for publicity. Yet both lacked the vital element of most bargaining proposals; the union-management relationship did not become intolerable if one party or the other sat on its hands. As a practical matter, the union had no power to turn them into genuine issues at the national level.

The weapons of combat. At the city level, pressures are available to force agreement on some terms. The most potent is the strike or lockout, but there are others which are also used. On the union side one of the most commonly applied has been the restriction on overtime. When negotiations are coming down to the line and agreement has not been reached on important issues, the local may require that no member work overtime. Placing what is essentially a job-shop operation into an inflexible hourly schedule is a penalty to management that is especially severe if business is good, and in group negotiations may have the effect of dampening the desire of some firms to "hold the line at any cost." In 1958 the Buffalo employers were hit with the overtime ban, and took immediate steps to have the prohibition declared an unfair labor practice under the refusal-to-bargain clause (8 [b] [3]) of the Taft-Hartley Act. They were successful, and in August 1959, the NLRB declared that Local 2 in Buffalo had engaged in an unfair labor practice, basing the decision specifically on an earlier decision in the *Insurance Agents International Union (Prudential Insurance Company)* case.[21] The following February this decision was overturned by the United States Supreme Court, and it would appear now that although an overtime ban during a contract with a no-strike clause is illegal, such a ban after the contract expires is unprotected activity, but is not a refusal to bargain *per se,* as the NLRB ruled. As Justice Brennan noted, "It may be that the tactics used here deserve condemnation, but this would not justify attempting to pour that condemnation into a vsesel not designed to hold it."

Another effective union tactic is at least a twin to the secondary boycott. Trade shops are particularly vulnerable to a boycott enforced by the membership in firms using trade shop plates or color separations. The application of such pressure since 1947 has been illegal, but the ALA has been notably successful in avoiding illegal acts to reach what appear to some to be illegal ends. Nowhere was this better expressed than by Ben Robinson, when answering questions from the floor in an employer's meeting. He was asked by a New York employer what the union would do if a plate from a non-union shop came in. Robinson answered as follows:

So far as the law permits, or so far as there are practical substitutes for the law, that plate won't run . . . Of course, that is because the plate is no good.[22]

Relatively few pressures are available to employers in most cities, most of them depending on skillful presentation of facts at the bargaining table or before the union membership in the shop. The union is clearly in a less advantageous position than the employers when bargaining in periods of depressed business activity, but this is a pressure of the sort most employers would be happy to do without. Key pressures are available to employers in three situations. The first occurs when an employer has shops in other areas. Here the transfer of work from one shop to another may serve the same purpose for the employer as the restriction on overtime does for the union — a timely warning of what may happen if the other party pushes too hard. The second situation is that in which the employer can secure a sufficient number of replacements either nonunion or from another union. An ALA strike in Miami was broken largely by the IPP&AU's ability to provide acceptable substitutes for the striking workers. Miami is, of course, somewhat unique, since elderly IPP&AU members can have jobs, serve their union, and prepare for retirement all at the same time. The third situation is not common except in some southern cities, where the union is not accepted as a permanent institution. Here an employer, with the same confidence in his ability to find "practical substitutes for the law" exhibited by the union in its use of secondary pressures, may initiate a practice of selective firing and layoff, intended to break the power of the union or at least to scare it into a more conciliatory frame of mind. A recent strike in Oklahoma City was broken by such methods. Needless to say, this strategy is not available to employers in a city where the union is strongly organized, and it has not been a factor in bargaining in the major centers since the late 1930's.

A substantial number of large employers are multiplant corporations, and the ALA has on occasion suffered defeats because of the ability of employers to shift work from shop to shop. To control this situation the international has urged locals which have one plant of a multiplant corporation in their jurisdiction to secure the following wording in their contracts, known as the chain shop clause:

> The Company agrees that its employees shall not be required to handle any work in this plant of the Company if in another lithographic plant of this Company in any part of the United States or Canada any Local of the Amalgamated Lithographers of America or the International is on strike or lock-out.

Similar clauses to cover the exchange of work between plants not under common ownership raise issues more appropriately considered in a later chapter. Not all locals have been successful in securing the protective language. Some have it in modified form; others, such as New York and Chicago, have been able to secure it with no material change. But a number of locals have had no success at all in securing the language, particularly in cities where there are no firms competing in the national market, and the local union is oriented to the problems of local competition. This is also true in one-shop locals, where employer bargaining power is typically greater than that enjoyed by employers in the larger cities with multi-employer bargaining.

The final weapon is the strike or the lockout. Both are expensive to the membership, the union, and the employers, and the strike weapon on occasion has been misused. The union went through a period in the postwar years when ill-advised strikes were permitted; it has already been said that one of them contributed to President Blackburn's defeat. But there are also unavoidable strikes. One in Los Angeles was typical; the parties were too far separated on issues, and the local had never proven itself on a picket line. The great value of a strike or lockout is in the damage threatened, not caused. If management believes that a strike cannot or will not be called by the union, at some time the union will be forced to put up or shut up, and this occurred in Los Angeles. In spite of this, it is probable that the majority of strikes in the period from 1955 to 1960 have been for recognition, rather than for the resolution of conflict at the bargaining table. The ALA does not have the reputation of being strike-happy, but it has called a sufficient number of strikes in recent years to convince many employers that most locals can and will sustain a strike when their interests require it.

Lockouts are far less common, in part because they are illegal except when used in a purely defensive manner. Canadian employers used the device successfully in combating the union's one-at-a-time tactics in 1949, and some local associations in the United States are said to have resorted to it for the same reasons, though no recent instance is known to the writer. The nature of the management group in most cities is such that cooperation to this extent is highly unlikely in the absence of a vigorous anti-union attitude among all employers and of a product market whose boundaries are contiguous with those of the employer group. As already noted, this is not the case in most cities.

The typical employers' group in lithography today finds it exceedingly difficult to secure member support for a strong line against the union. During the Boston negotiations the possibility of taking a city-wide strike was mentioned in passing, but none of the committee seriously considered the prospect. One member recounted an incident in a prior negotiation when the employers met together and unanimously agreed that the union demands were outrageous and should be resisted to a strike if necessary. After the meeting one employer approached the narrator and said, "Jeez, Al, we can't do that." Much the same apparently took place in Chicago during the thirty-five-hour-week negotiations.[23] When the negotiations got down to the wire, the union made it clear that some employers who were members of the employer association had informed the union that they would sign a thirty-five hour contract rather than take a strike, and the employer bargaining committee knew that a strike would leave a substantial number of large shops working on union terms. One trade association executive expressed himself forcibly on this problem:

Perhaps the greatest weakness in employer negotiations is a very puerile mental attitude. No matter how many times they have been disillusioned, employers by and large always approach negotiations with the conviction that all they have to do is explain "the facts of life" as they see them to their "boys" and they will gain support. As a consequence they will not consider the possibility of a strike six months or even 30 days in advance and make some sensible preparations. There is much they could do. But they do not. And when the union flexes its muscle they are entirely unprepared and have no option but to yield.[24]

There are also other explanations for employer reluctance to take a strike. Much lithographic work is now done on a supply contract. Accordingly, the costs of a strike have to be measured not only by the work lost during the strike but by the contracts lost which can be regained only over a period of months or even years. Among the smaller shops the existence of the buyer-supplier bond facilitates the immediate passing on of cost increases. Thus the situation is such that the firms in price competition cannot afford to take a strike, and the firms that can afford to don't need to. Although in the long run the ALA faces some serious challenges to its bargaining position, in the short run the pressures which it can exert to secure a favorable settlement are considerably more effective than those available to the typical employer group with which it negotiates.

Administering the Contract

Formally, contract administration is quite separate from the determination of contract terms. Yet one of the more curious aspects of the 1958 Boston negotiations was the overlap between bargaining over the terms of the contract and administering or interpreting it. In several cases — in particular when the union was seeking language to protect it from employers who in its view abused the "flexibility clauses" in the contract — the proposal was made that in order to avoid filling the contract with technical qualifications covering each eventuality, the problem could be left to the area grievance committee.

In at least two cases it was clear that the employer and union bargainers did not agree entirely on where justice lay in the incidents which provoked the union requests for restrictive clauses. Apparently both sides had faith that when all facts were known, a solution would be forthcoming. On another occasion already referred to, the ticklish question of complement, one of the decisions which helped break the impasse was the resolution of one of the differences on an individual shop basis, specifically leaving it out of the contract. This instance also illustrates the danger of allowing the terms of the union-management relationship to be settled in several different forums over an extended period of time. Well after the contract was signed, it was discovered that there was still a minor but very touchy difference of opinion over whether this decision, which was in fact a union concession, applied to one shop or several. To the considerable embarrassment of the two bargaining committees, this was discovered first at the meeting between all shop delegates and employers of the contract group, called to educate those who had not sat in on the bargaining over what the contract really meant.[25] The eventual resolution of the disagreement came through further meetings of the two bargaining committees.

Disputes over the meaning of the contract are settled through a formal grievance procedure which typically takes one of two forms. In both forms the first step is always within the shop; the second either requires a discussion between the local union president and the employer of the unsettled grievance, or a discussion between a committee of employers and union members. In all contracts reviewed, there is provision for arbitration as the final step. The Boston contract provides for the committee type of second step, and that city is used in the following description.

The switch to an area grievance committee came through employer initiative, in order to provide a formal method of achieving uniform contract administration. This was in no way objectionable to the overworked union president, and both parties continue to agree that the change has been desirable. The committee members, two from each side, are appointed anew for each grievance. No person from the shop involved may sit on the committee. The decisions of the area grievance committee are treated as arbitration awards:

Such settlement shall then become the uniform interpretation and application of that particular provision or provisions of the current agreement for all shops signatory to the agreement.

An additional device, to the writer's knowledge practiced only in Boston, gives the committee a preventive role:

From time to time a Joint Standing Committee [area grievance committee] may convene on call of either party to consider provisions of this Agreement which require case examples or interpretations to be sent in written bulletin form to all Employers and to all Shop Delegates. To assist in the identification of these provisions, copies of all written grievances filed by either party and decided at the plant level shall be sent to the Union and to the Graphic Arts Institute.[26]

This was only written into the contract in 1958, and its usefulness has not yet been adequately tested.

The employer negotiating committee in Boston has had some difficulty in convincing employers that the grievance procedure is also for their use. The point has been made strongly because of the employer committee's desire to enforce the proposition that *all* differences are to be settled by the peaceful procedures provided and not by slowdowns, quickies, or cessation of work while a delegation goes to discuss the ALA constitutional requirements with the management.

The use of arbitration is rare. For example, only a half-dozen grievances have reached the area grievance committee in Boston during the last two-year period. With one exception (an employer grievance) all were settled at that stage. The great majority of differences are settled informally either at the shop level, by personal discussion, or on the telephone with the Local 3 president.

Other formal relationships between employers and the union may be disposed of briefly. In most cities there is an apprentice committee which establishes training programs and is the forum for discussion of the manpower needs of the industry. In locals which have an employer-financed health and welfare program there is a board of trustees responsible for determining that the level of benefits to be paid does not exceed the funds available. It is not given the authority to determine what benefits will be paid. New York has an additional committee whose function is to consider the need for retraining journeymen presently in classifications being reduced or eliminated by the pressure of technological change.[27]

These committees arise in part from the internal structure of the union. There is a distinct tendency to avoid giving the union president the power to act for the union in these areas because of the traditional dislike of a strong executive. Moreover, the same argument holds for a group of employers who make up an association typically without centralized authority. The multiple contracts between employers and union members which these committees provide have a value of their own, quite apart from the assigned function of the committee. These nonnegotiating relationships are an important and useful part of the total union-management relationship.

Chapter **X** ✆

Results of Bargaining:
Distribution of Money

Lithographers have done remarkably well in the postwar period. A majority of them work a thirty-five-hour week, are covered by broad health insurance, life insurance, and pension programs, and they have progressed more rapidly on the wage front than most other printing industry crafts.

The Basic Lithographic Wage Contours[1]

A wage contour, as the term is used here, describes a group of firms or bargaining units whose wage rates are closely related. The relationship is enforced by similar product markets, similar labor markets, or a single labor organization. In the lithographic industry there is a basic wage contour which is linked through product markets and a single labor organization — the ALA. It is not linked through a single labor market because it is a national wage contour. But it does include a limited group of labor skills described in the workforce chapter as the high-skill job cluster. The high-skill cluster makes up the basic wage contour in lithography. It is a national wage contour and contains all the jobs for which rate comparisons are made among cities both by employers and union. There is a tendency for rates in the same region to be comparable and a tendency for rates in cities of similar size anywhere in the country to be comparable. Even where rates vary, the rate changes in all cities in any series of contract negotiations are likely to be similar, regardless of the size of the city or its regional location.

This basic wage contour is roughly contiguous with the national product market. The patterns are normally set in New York or some other large city, whose leading employers are most directly concerned with the problem of competition from other cities. The union then attempts to extend the substance of these bargains throughout the country in successive negotiations. The degree of union success in keeping other cities up with New York (and each other) is significant; as we noted earlier the key employers sell in a national market, and this national market unlike the local one is sensitive to price differentials.

Table 10 compares national wage differentials in three classifications over the period 1943–1958. The table is not really complex and deserves close attention.

Table 10.   Interlocal wage comparisons, 1943–1958.

City	Color photographer				Color stripper				2-color 58″ pressman			
	1943	1947	1952	1958	1943	1947	1952	1958	1943	1947	1952	1958
New York	$1.87	2.95	3.53	4.11	1.43	2.28	2.86	3.63	1.75	2.77	3.34	3.92
Per cent difference from Chicago	+7	+20	+20	+10	+4	+14	+14	+5	0	+12	+12	+9
Chicago	$1.75	2.46	2.97	3.75	1.38	2.00	2.51	3.44	1.75	2.46	2.97	3.59
San Francisco	$1.75	2.36	3.19	3.58	1.38	1.96	2.77	3.38	1.57	2.20	3.13	3.61
Per cent difference from Chicago	0	−4	+7	−5	0	−2	+10	−2	−10	−10	+5	+1
Boston[a]	$1.50	2.14	2.80	3.58	1.25	1.64	2.28	3.09	1.75	2.38	3.04	3.82
Per cent difference from Chicago	−14	−13	−6	−5	−9	−18	−9	−10	0	−3	+2	+6
St. Louis	$1.60	2.28	2.71	3.59	1.30	1.84	2.23	3.12	1.60	2.28	2.71	3.59
Per cent difference from Chicago	−9	−7	−8	−4	−6	−8	−11	−9	−9	−7	−8	0
Pittsburgh	$1.63	2.13	—[b]	3.65	1.50	1.93	—[b]	3.14	1.87	2.13[c]	—[b]	3.65
Per cent difference from Chicago	−6	−13		−2	+8	−4		−9	+7	−13		+2
Milwaukee	$1.73	1.99	2.81	3.57	1.21	1.63	2.41	3.16	1.50	1.87	2.73	3.58
Per cent difference from Chicago	−1	−19	−5	−5	−12	−18	−4	−8	−14	−23	−7	0

Sources: 1943 (except Boston), *Minimum Wage Scales* (March 31, 1943), prepared by Vice-President Robert Bruck; 1943 Boston Submission to Dept. of Labor and Industries, Massachusetts (June 1, 1944); 1952–*ALA 1952 Wage Survey;* 1958–*ALA Wage Scales and Working Conditions* (January 1958). All save the *1952 Survey* have been made available through the courtesy of Arthur Willis, former president, Local 3.

[a] 1943 figures as of June 1, 1944.
[b] Not available.
[c] Rate is for 2-color 55″ pressman.

Although it seems clear that there can be substantial differentials among cities without any noticeable shortrun effect on the industry, the evidence indicates that New York rates for a time may have exceeded this differential.[2] In the 1958 negotiations Local 1 agreed to take a bye on wages for the first year of a two-year contract as a direct result of an impressive statistical presentation by the Metropolitan Lithographers Association. The association used Census data to show, quite accurately, that New York was losing its relative position in the industry, evidence from employers to show that jobs were going to outside firms through underquoting, a listing of non-New York lithographers having sales offices in New York to show the extensive outside competition for New York customers, and prevailing wage data to show that labor costs were the key factor.[3] Local 1 was impressed, and President Swayduck accepted the task of convincing the membership that higher scales with no overtime and possibly short-time were a poorer investment than a continuation of existing scales while the rest of the country caught up with New York.

The idea of urging the rest of the country to catch up with New York has been a key feature of international policy in the postwar period, though for political reasons it is not always explicitly stated. Local 1 New York plays continual variations on the theme, sometimes aiming at one-shop locals such as those in Poughkeepsie, New York, Bennington, Vermont, and Ashland, Ohio; sometimes at the large cities like Chicago, Philadelphia, and the Ohio group. At times when the international is not giving due emphasis to this aspect of wage policy, Local 1 goes to work on the international, as it did when William Riehl was president and more recently when George Canary seemed not to support the policy with sufficient vigor.[4] In addressing a group of New York employers in November 1958, Ed Swayduck told them that they were going to get some of the work back that they had developed; the ALA was going to "do a better job negotiating out in the cornfields."[5]

New York does not always push this policy alone. Some leaders in other large locals realize that the ability of New York to establish a pattern which may then be applied nationally is a valuable aid in their own negotiations, and that hamstringing New York by not getting the rates in line "out in the cornfields" hurts their own bargaining power. Other leaders resent the attitude of confident superiority which has been a trade mark of Local 1 leaders since 1882, and may combine this resentment with a belief that they should not push their employers too hard, because of local wage comparisons.

The local printing wage contour. The basic contour also contains jobs which in other shops are part of the low-skill job cluster, dominated by lesser-skilled workers who work on the offset duplicators, small offset presses, or on the simplified offset platemaking equipment. These workers seldom produce for the national market, and their employers are not locally secure from wage competition. The strongest element which ties these workers to a wage contour is competition in the local product market, and therefore they are more firmly attached to the local printing wage contour than to the national lithographic contour.

This second group of local contours is made up of workers who account for much of the growth of the lithographic workforce in the past two decades. The patterns of wages, hours, and working conditions are not set only or even primarily by the ALA, but by negotiations between the IPP&AU dominated by letterpress workers, and the local employer associations dominated by the letterpress employers. Rates on offset equipment tend to follow rates on letterpress equipment. This is understandable since the workforce is interchangeable to some degree, the same union bargains for both groups of workers, and the employers of both groups serve the same product market. When these workers are represented by the ALA, the pressure to keep in line with the IPP&AU locally is much stronger than the pressure to keep in line with the ALA in some other city. This is well illustrated by comparing local contract rates for the ALA and the IPP&AU in a half-dozen smaller cities, with the Chicago rates for the same job to provide a national benchmark.

Though somewhat apart from the main argument presented here, it is

Table 11.   ALA and IPP&AU hourly rates on lithographic jobs, October 1957. (in dollars)

City	Black and white cameraman		Feeder		17″ x 22″ pressman		2-color 22″ x 34″ pressman		Work week hours	
	ALA	IPP& AU	ALA	IPP& AU	ALA	IPP& AU	ALA	IPP& AU	ALA	IPP& AU
Chicago	3.44	—	2.46	—	3.16	—	3.59	—	35	—
Buffalo	3.23	2.97	2.45	2.67	3.05	2.88	3.63	no rate	35	37.5
Atlanta	2.94	2.94	1.87	1.88	2.77	2.77	3.07	3.07	37.5	37.5
Boston	3.32	no rate	2.45[a]	2.54[a]	2.70	2.67	3.71	2.85	35	37.5
Des Moines	2.98	2.33	2.10	2.33	2.72	2.71	3.12	3.12	36¼	37.5
Kansas City	3.22 (3.53)[b]	3.10 (3.40)[b]	2.37	2.38	2.95	3.00	3.44	no rate	35	37.5
Wilmington	3.42	2.75	2.27	no rate	2.87	2.81	3.78	2.86	35	37.5
Seattle	3.31	3.29	2.35	2.64	3.31	3.29	3.57	3.50	35	35

Sources: IPP&AU rates: *PIA Lithographic Manual*, IPP&AU Contracts Section; ALA rates: *NAPL Labor Relations Manual*.

[a] Feeder on large 1-color equipment (over 65″).

[b] Color cameraman rate.

worthwhile to note the sharp contrast which the feeder rate comparison presents to the others. The IPP&AU is the International Printing Pressmen and Assistants' Union. The press assistants have maintained their separateness from the other crafts far more successfully in the IPP&AU than have the feeders in the ALA. There are several locals for assistants only in the IPP&AU, and some even negotiate separate contracts from the pressmen. It appears that low-skill segments of a craft union may fight wisely when they fight to preserve their separate identity.

Table 11 illustrates that wage levels in shops serving a local market are closely related to the wage levels in that local market. One would therefore expect that nationally the average wage in smaller shops (measured by employment) would be somewhat lower than that in large firms. In fact, the Census data indicate that this differential is much greater than the rate differentials alone would lead one to expect. In the smallest shops (one to three workers) average annual wages in 1954 were $2740, but in shops having over fifty workers the average wage was between $4400 and $4500.[6]

One of the most significant factors is that many small firms are unorganized and pay much less than union rates, whether they be rates negotiated by the letterpress or lithographic union. The buyer-supplier bond that permits printing firms to hold on to customers in the face of substantial price differentials protects the union rates in the shortrun; and, as the small firm grows, it becomes an object of the union organizers' attention, and rates either go up to keep the union out or go up because the union comes in. But during this period of growth, and for those shops which remain small, the relatively much lower rates act as a definite check on union wage levels.

This has been an historic problem for the letterpress unions, but a relatively recent one for the ALA. As a result the ALA's negotiating policy in the early postwar period ignored the pressure from the local wage contour. Perhaps the best example of this is the handling of the Multilith rate.

When the Multilith was brought into combination shops and occasionally into straight lithographic plants organized by the amalgamated, it was often treated as an offset press, which in fact it is. The Multilith operator was called a pressman, and the rate established was often that for the 17″ x 22″ offset press. Treating the Multilith operator as a pressman tied the wage rate to the pressroom rates, and, as these moved upward, the Multilith rate went up in the same degree. Among the large locals, only San Francisco and St. Louis provided for a separate rate in 1943, and New York did not do so until after 1948.

More than a separate rate was necessary, however, if any small lithographic equipment were to operate profitably under ALA group contracts. It was necessary to apply different standards for appropriate increases from those relevant in the case of the skilled workers because control of the labor supply was impossible, and entry costs to the product market were remarkably low. Few lithographic firms cared to take a stand on the issue of Multilith rates, and those which did could expect little support from the rest of the employer group. In many cases the rates went up, and the Multiliths went out of the industry into the offices of the lithographer's customers. With the Multilith went a lot of the "bread and butter" work, the simply prepared, one-color, low-quality work on which low price was essential. George Strebel, the executive vice-president of the Printing Industries Association of Western New York has estimated that in the Niagara Frontier area alone there are about three thousand Multilith installations in shops and offices, doing much work which in former years would have been given to a commercial printer or lithographer.[7]

Multiliths have been purchased by the lithographer's competitors as well as his customers. Small, usually nonunion print shops can easily find a man at two dollars an hour who can handle a Multilith and is happy to work a forty-hour work week with time-and-a-half for overtime. By giving him the same take-home pay that he would receive in an organized plant the employer can get nearly a third more hours of productive labor from him.

Differences in hours, overtime provisions, and miscellaneous fringes can permit an employer to match hourly wages and still enjoy a substantial cost advantage over his unionized fellow employers. An employer in Boston recently set up an independent company (nonunion) so that the work handled most effectively on small presses could be done at a profit.[8] In the Boston contract the Multilith rate is $2.59 an hour, 9 percent under the 17″ x 22″ rate. Some locals have been able to split the rate more effectively than Boston. The New York rate is 23 per cent under the 17″ x 22″, the Buffalo rate is 25 per cent under, Baltimore, 30 per cent, and so forth.[9]

Loss of the Multilith work is not a serious problem either to the union or the great majority of employers who negotiate with the amalgamated. But it illustrates a problem which is growing in importance because the ALA is losing control over the supply of labor in the next higher range of skill classifications, the one-color and small two-color pressman group. As the ALA control over the supply of labor diminishes in ever higher skill classifications, so does its bargaining power. Here, however, a qualification need be entered. As the classification under discussion moves up the ladder of skills, so ease of entry to the product market diminishes because of the market structure and entry costs. Moreover, the ALA need be less and less concerned with open-shop rates and worry only about IPP&AU rates. This means that less flexibility is required to keep the rate structure in line with the local market. Although letterpress rates have not moved up as fast as those in lithography, the data presented earlier show that the disparity in most locals is not large.

Other industry wage contours. The local printing wage contour presents a long-term challenge to the bargaining power of the ALA because it threatens the continued job opportunity of the membership in the lower skill groups. At the other end of the skills spectrum, the ALA is in a similar difficulty because of the extension of the process into other industries.

The metal-decorating industry is dominated by four companies whose plants more or less blanket the country, metal decorating being an operation carried out most economically close to the final market.[10] This development is relatively recent, but the integration of lithographic operations into the canmaking process is not. Captive operations have existed for a long period in the industry, long enough for the ALA to be the traditional representative in such plants when the spread of unionization in the late 1930's established plant-wide units for the majority of the workforce. Metal decorating is a well-organized operation, with minimal product market competition from shops not organized by the union.

Publishing houses present a contrast to this. Nearly all large publishing houses used letterpress in the 1930's, and such establishments were traditionally organized by the letterpress unions. With the introduction of the offset process to such firms, the IPP&AU typically remained as the bargaining agent, and it is this sort of operation which is the most important provider of highly skilled IPP&AU lithographers. Rate differentials between the ALA- and IPP&AU-organized shops in such cases are not substantial. The four cities shown in Table 12 illustrate this. Chicago rates are again shown to facilitate comparison.

Another type of captive plant is the lithographic department of an integrated paper box plant. This is a rapidly expanding operation, and its expansion is typically in plants that are organized by one of the paper workers' unions, or by the IPP&AU specialty workers. Both wages and hours for lithographers in any of these unions are usually well below ALA conditions, or those of IPP&AU members in the printing industry. A 1956 contract for a paper company in the Boston metropolitan area carried a rate

Table 12. *ALA and IPP&AU hourly rates, four cities, October 1957.*
(*in dollars*)

City	2-color 58" pressman		4-color 68" pressman		Color cameraman		Color stripper		Photo-composing machine operator	
	ALA	IPP& AU	ALA	IPP& AU	ALA	IPP& AU	ALA	IPP& AU	ALA	IPP& AU
Chicago	3.75	—	4.14	—	3.75 (3.44)[a]	—	3.44 (3.44)[a]	—	3.26	—
Detroit	3.40	3.36	3.99	3.96	3.41	no rate	3.41	no rate	3.29	no rate
Philadelphia	3.88	3.31	4.09	3.71	3.91 (3.56)[a]	no rate (3.16)[a]	3.27 (3.20)[a]	no rate (2.90)[a]	3.38	2.97
St. Louis	3.59	3.52	4.36	4.27	3.59	3.52	3.12	3.06	3.12	3.06
Washington	3.35	3.28	no rate	no rate	3.50 (2.91)[a]	no rate (2.96)[a]	3.04	2.96	3.04	2.96

Sources: ALA rates: *NAPL Labor Relations Manual;* IPP&AU rates: *PIA Lithographic Manual,* IPP&AU contract section.

[a] Rate for black and white/halftone classifications.

of $1.68 for an offset pressman, size of press unspecified. In the same year the lowest pressman rate in the ALA *Boston Group Contract* was $2.38. In 1958 the Sutherland Paper Company plant in Kalamazoo, Michigan, had rates well below anything in Detroit or Chicago, as Table 13 illustrates:

Table 13. *Lithographic rate comparisons, 1958.*
(*in dollars*)

Category	Kalamazoo[1] (UPP)	Detroit[2] ALA	Chicago[2] ALA
5-color pressman	3.13	4.09[a]	4.14[a]
5-color second pressman	2.48	3.40[a]	3.44[a]
Platemaker	2.60	3.29	3.44

Sources: (1) In the Matter of Sutherland Paper Company, NLRB 7-RC 3717, 1958, *Brief of the ALA,* pp. 40–41; (2) ALA *Scale Survey* (January 1958).

[a] Rates are for 4-color presses.

These plants under contract to the paper workers' unions are under constant attack by the ALA. Ace Folding Box, Sutherland Paper, Fibreboard Company, and other such firms account for a substantial portion of all large multicolor press purchases in recent years and compete directly with companies whose lithographers are organized by the ALA. When low wage rates, a forty-hour week, and poorer conditions in relation to overtime, vacation, and holidays are added to the inherent economic advantages of an integrated operation, it is clear that the rapidly expanding market for lithographed cartons will be lost to ALA-organized plants unless wage competition can be eliminated.

Two somewhat contradictory facts should be mentioned. In general, lithographers in the integrated box plants require fewer skills than those in color houses. This is because the variety of products on which they work is more limited than in a regular lithographic establishment. It has elsewhere been noted that the degree of craft skill is closely related to the range of products which can be handled. The Sutherland Paper Company started men on five-color 76-inch presses after twelve weeks of training, which is measurably less than the eight years (four as apprentice feeder and four as apprentice pressman) required by the ALA constitution.[11] On the other hand, Sutherland did not expect to handle all jobs itself and did continue to use trade platemakers although it had its own platemaking shop. Other employers have said, in discussing this aspect of competition in the industry, that it will be some time before lithographic firms feel the competition from such firms as Sutherland. The output of the latter will indeed expand rapidly, but this will be in an expanding market, and these firms will in any event not be competent to turn out a quality of work comparable to the lithographers for at least three to five years.

The IPP&AU is also seeking control of the press in these plants but has adopted an entirely different strategy. The pressmen are quite willing to organize industrially, and their specialty workers' branch is in fact a variant of the United Mine Workers' District 50, except that it concentrates on plants having some sort of printing operation. All workers in the plant are taken in, and no apparent attempt is made to tie press rates to the commercial printing wage contour. Rates, hours, and working conditions in box plants organized by the IPP&AU are competitive with those in plants organized by the two paper unions, the IPP&AU believing that any attempt to do otherwise would be tilting with windmills.

The letterpress unions have long accepted the fact that the different economic conditions in newspaper plants on the one hand and commercial printing and publishing houses on the other require different approaches to bargaining, and separate contracts in the same city for these two employer groups have been the rule since the turn of the century. A precedent was thus available when box plant employers insisted that their industry was different, and that a successful organizing drive would only be successful in driving printing operations out of the firm if conditions prevailing in the commercial printing industry were forced on them. Believing this to be true, the IPP&AU leadership is in full agreement with George Meany that the ALA approach simply illustrates a case of "very stupid leadership."

It must be remembered that the ALA and the IPP&AU leadership do not merely have disagreements; they actively dislike each other and have for a half-century. Having chosen to go the plant-wide route with their specialty workers, the IPP&AU is no longer free under NLRB rules to opt for a craft unit in the folding paper box industry. Thus the ALA is the only union able to seek and which *is* seeking such a unit.

In a sense, the ALA has a period of grace in which to extend its domain (or at least its working conditions) to lithographic workers in integrated box plants. The prospects for doing so are probably less favorable than they were when the lithographers in can plants were organized because now there is a rival union in the box plants. Prospects are probably more favorable than in the publishing houses since the rival is not a craft union. But it is clear that if the ALA is unable either to force rates up through organizing or through fear of organizing, the basic lithographic wage contour will be weakened at the top of the skills ladder just as it is at the bottom. Again, the ALA has ample bargaining power in the short term because product market mechanisms are such that long lags are introduced in the process of working out new market arrangements. But in the long term their power in the labor market is meaningless if the buyers in the product market adopt a new set of suppliers.

Trade shops. Other workers are also subject to pressures somewhat different from those in the basic contour. These are the preparatory craftsmen in the lithographic trade shops. When rates are the same in trade shops and integrated lithographic establishments, there is some tendency for work to shift to the trade shops because of the more efficient use of skills and equipment possible there. An additional reason is that there is greater difficulty in enforcing working rules with respect to overtime, piecework, subcontracting, and other control devices in a trade shop than in an integrated establishment. Workers in trade shops are not to be considered in a different wage contour, however. The fact that they *are* in the same contour creates the difficulty.

From the union point of view it is possible to do no more than fight a holding action against trade shops. There is a sufficient number of nonunion trade shops to force the union to recognize that too stiff a policy toward those organized would merely produce more nonunion shops since entry is remarkably easy. Actual wage rates have to be kept in close relation to rates for identical skills in integrated plants, or the labor market is split, reducing the mobility on which the union's day-to-day effectiveness in the labor market largely depends. The device adopted is one to warm the heart of an economist; the union encourages premium payments to individual workers (hourly rates in excess of negotiated scale rates) so that actual wage payments are higher in trade shops, and wage flexibility is retained, while trade shops and integrated shops continue to operate under the same contract and wage scale. To take one example, employers in Chicago have been notably successful in resisting the introduction of premiums in the majority of integrated lithographic shops. One employer in discussing this admitted that on occasion he loses a good dot etcher to the supposedly greener pastures of a trade shop. If the worker can stand the pace he admittedly profits by it.[12] By the same token, workers who cannot keep it up and do not enjoy the uncertainty of continued employment may well prefer to go back to an integrated shop.

Premiums also act to some degree as a replacement for piece rates, which as noted are more likely to be attractive to trade shop employers than to others. Although the practice of paying premium rates does not provide as close a relationship of the wage bill to output as that of paying piece rates, it reduces any desire of management to experiment with piece rates.

It is not easy to find meaningful generalizations that will apply to ninety-odd local bargaining situations, and the handling of the trade shop question is no exception. For example, Toledo is predominantly a trade shop town; Detroit, just up the lake sixty miles, is not. Yet in 1947 Detroit had greater premium payments in the preparatory classifications than Toledo, although in 1956 the reverse was true.[13] In larger cities where trade shops are well established, premium payments tend to be consistently higher in such shops than those paid in integrated shops.

Wage Variations

Premium payments. The practice of paying above scale to individual workers is not limited to trade shops, nor is it a recent development. The desire to control premium payments was evident in a 1902 employer association rule that no member firm could hire a workman leaving another member firm at a rate higher than that received from his former employer.[14] Typical premiums paid in 1956 are illustrated in Table 14. The locals have been chosen to illustrate four different situations. New York is a strong local with high labor mobility; Chicago is a strong local with less intralocal mobility; Rochester is a weaker local, 50 per cent of whose members are general workers; and Racine is a one-shop local.

Table 14. Premium payments over scale rates, 1956.

Category	New York	Chicago	Rochester	Racine
Dot etcher/process artist:				
1. Scale rate	$3.84	3.46	3.60	3.37
2. Prevailing rate	$4.90	4.09	3.94	3.62
3. % journeymen over scale	n.d.	86	50	90
Color photographer:				
1. Scale rate	$4.11	3.60	3.60	3.60
2. Prevailing rate	$4.85	4.17	3.85	3.84
3. % journeymen over scale	n.d.	75	65	100
4-Color pressmen:[a]				
1. Scale rate	$4.34	4.00	3.94	4.03[c]
2. Prevailing rate	$4.51	4.03	4.26	4.06[c]
3. % journeymen over scale	n.d.	20	100	—[c]
4-Color feeder:[b]				
1. Scale rate	$2.87	2.69	2.59	—[d]
2. Prevailing rate	$2.93	2.69	2.59	—[d]
3. % journeymen over scale	n.d.	none	none	—[d]

Source: ALA, *Prevailing Wage Survey, 1956*, internal record of the union. Approximately two-thirds of all journeymen returned questionnaires.

[a] Data for 4-color 68″ first pressman.

[b] Data for 4-color 68″ first feeder or operator.

[c] Data for one 4-color 72″ first pressman.

[d] No journeyman in the classification responded.

Premiums primarily reflect skill differentials, and therefore vary widely among workers in the same classification. Evidence of this is particularly striking in the *1947 Survey* made by the union. To take several examples from the New York local, 202 dot etchers were receiving an average of 30 per cent above scale although 11 were at scale and 9 below.[15] In another classification averaging a 5.5 per cent premium, nearly 20 per cent were below scale and over 20 per cent on scale. The *1947 Survey* alerted the union to the number of below-scale members and in a number of locals led to investigation of the number of apprentices working as journeymen, the number of journeymen in contract shops actually receiving below-scale rates, and so forth.[16]

Table 14 also points up the value of premiums in providing wage flexibility. Once wage relations between jobs are established, the established differentials tend to be viewed as "fair" or "reasonable" in the eyes of workers, so that changes in them, particularly reversals, are vigorously resisted. When it is recognized that pay differentials are often viewed as the best evidence of status differentials, this resistance is not irrational. Yet such attitudes do lead to a distribution of labor which is less than perfect, and it is from this point of view that the wage data for New York in the table are particularly gratifying. Although the scale rate for the three skilled classifications places the pressman at the top and the dot etcher at the bottom, the actual payments precisely reverse the relationship.

Premiums cannot exist readily unless the supply of labor is limited. It is not possible to limit the supply of labor easily in classifications in which little skill is required, and consequently such classifications as feeder operator, offset duplicator operator, and similar low-skill jobs seldom receive premiums. Premiums are also uncommon in dying skills. Even in the *1947 Survey* premiums had been largely eliminated for the transferrer and plate-graining machine operator, and from 1947 to 1952 premiums decreased noticeably for the photo-composing machine operator. Photo-composing is not a dying skill, since plates still have to be exposed. But the introduction of long-life plates had reduced the demand for platemakers more rapidly than the halt in making apprentices and retraining platemakers in other classifications had reduced their number.

In common with any other shop practice, in time the payment of premiums takes on traditional justification quite apart from the original reasons for its introduction. Thus in some locals a craftsman who does not receive a premium above the scale is considered barely more than incompetent. This has been the traditional view of the ALA toward wage scales and was part of the reason for the union's reluctance to sign agreements during the 1920's. Unified employer groups were in many locations able to agree on and enforce the rule that no above-scale payments would be made. During the 1930's the growth of interregional wage differentials and the increase in unskilled workers in the union caused a shift in union attitude toward negotiated scale rates, and it is now more common for locals to direct considerable bargaining attention to them.

In general the ideal situation for the union is to use across-the-board increases to grant everyone some increase, and larger scale increases to soak up the premium spread which has developed since the last negotiations. It is easier to build premiums when the scale is fairly close to the going premium than when the typical hiring rate for a skilled craftsman is from 40 to 50 per cent above the scale. Moreover, a relatively low scale is a serious bargaining disadvantage to a sister local in an adjoining area.

Prior to the war, wage levels were stable, and premium payments were probably less common than below-scale payments. But from 1945 on, wage movements went steadily upward. Scales moved up unsteadily, jumping to new levels only when a new contract was negotiated. Premiums followed a steady upward trend, so that an across-the-board increase that was added to scales would at each successive negotiation leave premiums starting from levels further removed from scale rates. In some respects, this steady upward movement of wage rates is similar to the "wage drift" which has been noted in national wage patterns.[17] In a number of cities the practice grew of occasionally raising scales by more than the across-the-board increase, so that skill differentials did not get completely out of line. For example, in the 1956 New York contract, scales over the two-year period were raised $12 per week, although across-the-board increases were only $8 per week, plus any amount necessary to keep workers on scale.[18] General workers received $5 both on scale and across-the-board. In 1958 Cleveland raised scales by $7 per week with an across-the-board increase of only half that much, and Los Angeles had earlier in the year done much the same. Los Angeles was criticized by the president of the San Francisco local for selling its premium men short, although the fact that a week-long strike was necessary to close the bargain suggests that the wage gain was substantial.[19] There are political overtones in this criticism that cannot be separated from its economic justification. San Francisco President Brandenburg, as the incumbent west coast international councillor, had been defeated for this post by Los Angeles President Brandt. San Francisco's $12 across-the-board settlement had been topped by Los Angeles' $13.25 scale increase. These settlements, coming just before the election, no doubt played a role in Brandenburg's defeat.

Skill differentials. Skill differentials in all trades have closed up in the war and postwar periods, and lithography is no exception. In the prewar period and up to about 1943 it was not uncommon for a feeder to receive half as much as the pressman on the same press. Differences have not been so acute in the IPP&AU, where the press assistants have retained their separate identity to a greater degree than have their counterparts in the ALA. There is more overlapping on the rates of assistants on large presses with the rates of pressmen on platens today than there was fifteen years ago, but in general the pattern has been a slight increase in money differentials and a substantial decline in percentage differentials.[20] With the lithographers' union the case is somewhat different. Percentage differentials have been cut to a third, but there still has been room for some increase in money differentials. Premium payments to higher-skill groups have the effect of counteracting the compression in skill differentials, but this is only true for the higher classifications. And, as already noted, the practice of making scale raises greater than across-the-board increases has the effect of moving people at scale up by a greater absolute amount than premium men above scale.

Table 15. Compression of skill differentials, 1943–1958 (New York).[a]

	1943		1958		Increase 1943–58	
Category	Rate	Above low rate (in per cent)	Rate	Above low rate (in per cent)	Absolute	Per cent
Feeder, 17″ x 22″	$ .75	—%	$2.66	—%	$1.81	240%
1st feeder, large 4-color	1.03	37	3.01	13	1.98	192
Pressman, 17″ x 22″	1.10	47	3.29	24	2.19	199
Stripper, color	1.42	90	3.63	36	2.21	156
Dot etcher	1.70	126	3.84	44	2.14	126
Color photographer	1.88	150	4.11	55	2.23	117
1st pressman, large 4-color	2.02	270	4.48	68	2.46	121

Sources: ALA, *1943 Wage Scale Survey;* ALA, *1958 Wage Scale Survey.*
[a] These are *scale* rates, which are not necessarily the actual wages paid.

Relative wage changes: lithography and letterpress. There is no ready way of showing the relatively more rapid increase in wages among skilled workers in lithography than in letterpress. No one to the writer's knowledge disputes the fact, however, that, at least from 1941 to about 1955, this difference in increase occurred. Testimony to this effect is even available from President Berry of the printing pressmen. In 1940 he told the IPP&AU convention that lithographers' rates "are much lower than the rates paid to our union" [21] but in 1948 the emphasis had changed. "In the beginning and to some extent now, offset printing could be produced more cheaply than letterpress printing due in the main to low rates of compensation and inferior working conditions of many of the offset printing pressmen on the continent." [22]

Statistics are more difficult to provide. Census data require extensive interpretation, so extensive that the differential rate of wage increase which can be shown is less than the probable error introduced by the assumptions. BLS *Hours and Earnings* data are not helpful for the same reason. Negotiated wage scales for approximately equivalent classifications should in theory show this trend, and to some extent they do. However, it has already been noted that the ALA has been remarkably relaxed about the enforce-

Table 16. *Comparison of letterpress and lithographic hourly rates in Chicago, 1943–1959.* [a]
(in dollars)

Date	Dot etcher	Photo-engraver	Stripper (color)	Machine compositor	Platemaker (photo-composing only)	Electrotyper	22″ x 34″ pressman	Platen pressman	4-color 68″ pressman	10-color perfecting pressman
April 1943	1.63	—	1.38	—	1.25	—	1.25	—	2.06	—
July 1943	—	1.64	—	1.49	—	1.77	—	1.25	—	1.95
April 1947 (prevailing)	2.30 (2.47)	—	2.00 (2.08)	—	1.99 (1.98)	—	2.00 (2.02)	—	2.84 (2.84)	—
Jan. 1948	—	2.37	—	2.33	—	2.50	—	2.16	—	2.79
Jan. 1952 (prevailing)	2.82 (3.18)	—	2.51 (2.67)	—	2.67 (2.72)	—	2.51 (2.56)	—	3.35 (3.61)	—
July 1952	—	3.36	—	2.84	—	3.16	—	2.63	—	3.27
July 1957	3.59	4.03	3.44	3.35	3.26	3.50	3.28	3.08	4.14	3.57

Sources: Letterpress rates 1943–52: BLS, *Union Wages and Hours: Printing Industry*, 1943—Bull. 820, 1948—Bull. 979, 1952—Bull. 1134, 1957—Bull. 1128; Lithographic rates: 1943—ALA *Minimum Wage Scale Survey*, 1947—ALA *Prevailing Wage Survey*, 1952—ALA, *Prevailing Wage Survey*, 1957—*Chicago Group Contract.*

[a] Weekly scale rates would show a greater relative gain for letterpress trades, since in general they have a longer work week. Weekly take-home comparisons would show a greater relative gain for lithographic trades, since actual hours worked vary only slightly, and more of the lithographers' hours are paid at overtime rates. See the following chapter for a more detailed discussion of hours.

ment of wage scales. A preceding section has described the current prevalence of "premiums," straight-time hourly payments in excess of the scale. This practice causes negotiated wage scales to understate wage levels. Premiums were not common in the immediate prewar period — quite the contrary. In New York the negotiated wage scales did not change from May 1939, to March 1943, a period in which there was a series of increases of about $9 for workers above $35 and $10.50 for those below.[23] This represents from a 10 to 25 per cent increase. In addition, in 1941, after a $5 increase for those below $35 and a $3 increase for those above had been procured, a clause was negotiated for a $3 increase for those below scale, and an additional $2 each six months as necessary to bring them up to scale. Clearly, the 1939 contract rates had as little relation to actual wage payments as the current contract rates, except that they then overstated whereas now they understate the actual wage levels. For this reason the natural choice of New York has been discarded, and Chicago is used to illustrate the trend in rates. Inasmuch as the average of union scales in the letterpress industry, weighted by the number of workers affected, put Chicago under New York in 1942, over New York in 1957, and because present ALA scales in Chicago are under those of New York, the choice is not one calculated to overstate the degree of spread in the two groups of rates. The figures are given in Table 16. It will be apparent that in some rate comparisons the ALA has lost ground. This is clearly evident in the dot etcher photo-engraver comparison. In others it has held even. The typical comparison, however, shows a substantial pulling-ahead in the lithographic classifications with respect to those which are approximately equivalent in letterpress.

Fringe Benefits

In an industry of small firms a number of the normal fringes can be reduced in cost if several firms act as a unit in providing the coverage. Group life insurance was first established by the lithographers' union as a benefit entirely separate from collective bargaining. Known as the "mortuary benefit," it provided a five-hundred-dollar payment during the union's early history, which was raised to one thousand dollars in the early 1920's. In order to wean workers away from the union, the NAEL after the 1906 strike urged members to establish their own insurance plans,[24] and after the 1922 strike it set up a group life and disability insurance plan in which members who did not have their own plans could participate.[25] The nature of the benefits is not known, but coverage for life insurance was considerable, including more than 10,000 employees in 1923. Some firms also had some form of health insurance, though this may have been little more than paid sick leave.

In 1947 the first negotiated health insurance program was established in New York, and by September 1949, over 65 per cent of the ALA members were covered by negotiated group health and insurance programs.[26] By 1957, according to union estimates, such programs covered 99.2 per cent of the membership.[27]

Typical coverage may be illustrated by that provided in the Boston plan in 1959:

SCHEDULE OF INSURANCE[28]

1. Life Insurance (double indemnity) $2000
2. Weekly Accident and Sickness Benefit
   a) Accident: beginning first day   (13 weeks at 50% of
   b) Sickness: beginning eighth day   weekly earnings not
   to exceed a maximum of $50)
3. Medical Expense Insurance*
   Office Treatment .... per call: $3
   Treatment other than at the Physician's
   Office .... per call: $5
   Diagnostic X-ray and Laboratory Fee $50
4. Hospital Expense Insurance*
   Maximum Daily Benefit $18
   Maximum Additional Benefits $180
   Maximum Maternity Benefit $120
5. Surgical Expense Insurance* $300
6. Major Medical Expense Insurance $5000

   * Coverage extends to dependents. Medical Expense Insurance is limited to in-hospital treatment, $3 per day with a $93 maximum.

The administration is handled by the ALA Local 3 office, formally under a separate administrative entity, the Metropolitan Boston Lithographers' Insurance Trust. There are six trustees — three union and three employers — but their authority is specifically limited by the collective agreement:

A Board of Trustees . . . shall . . . administer the affairs of the fund except that the Union shall have the full power to designate the kind, source and manner of purchase of such insurance . . . This insurance to be purchased from a nationally recognized company or companies.[29]

The Boston employer contribution is currently $3.25 per employee per week, somewhat under the typical contribution of about $3.50. New York far exceeds this with a contribution of $5.50 per employee. New York provides in addition to the normal benefits an eye-care clinic, and extends benefits to dependents, and to retired and unemployed workers and their dependents.

Pensions.[30] Pensions, unlike welfare programs, can achieve considerable economies of scale even when participating members are widely scattered. New York again was the first to work out a multiemployer pension arrangement, although, with pensions, the administration is completely by the union. As such, a check-off of pension deductions is illegal, and the device used is that a separate check is made out to each employee, endorsed by him to the pension fund trustees, and then transmitted to the trustees by the employer.

The New York plan was not feasible for small locals, and in 1950 a union-wide program was established. Called the Inter-Local Pension Fund, the administrative machinery (including the two-check system) is patterned after New York's plan, and now covers eleven thousand members in twenty-seven locals. Integrating existing pension plans is in practice nearly impossible, and the New York local, as well as several others which have their own plan, is outside the inter-local fund. These plans, including New York's seven thousand, cover an additional eleven thousand ALA members.

The trustees of the inter-local fund are the presidents of the seven largest participating locals. The pension funds are invested directly by them, under the advice of a trust company, which in the case of the inter-local fund also maintains the individual account records. The trustees also have the responsibility of determining the type of benefit and the pension pay-out rate, which is presently at $2.50 per month per year of service (plus a small prior service credit). There is also a withdrawal right (not fully vested), and a benefit for death prior to retirement of $1500. The outstanding feature of the pension program is its flexibility. Pension levels are raised whenever the experience of the fund shows that it is possible, and, in view of the initial conservatism in the establishing of the plan, it is not remarkable that pension levels have been raised considerably. In 1950 the guaranteed pension rate was $1.36 per month per year of service; it has since risen in three adjustments to $2.50 with apparently good prospects for increasing still further.

The trustees have taken their responsibilities very seriously, to the extent of being unwilling to help "sell" membership in the inter-local fund to locals which are considering the establishment of a pension program. Friction has arisen at this point, and they have been criticized for not showing a shade more trade union enthusiasm and fraternal concern for the welfare of the uncovered membership. This, in addition to the conservative setting of pension benefits and the quite rigorous standards for admission to the fund, has on occasion made it difficult for local leaders to convince the membership that joining the inter-local fund was a wiser move than buying pensions from a private carrier. Another type of difficulty has also arisen. A number of locals, which have secured the necessary language in the contract and negotiated an increase on the assumption that a pension program would be initiated, have encountered great difficulty in prying the $2.50 loose from the membership. Boston, for example, negotiated language in 1954 but only joined in 1959. During the 1958 Boston negotiations the union negotiators were perfectly frank in admitting the difficulty with the members, and on one occasion even suggested putting the $2.50 in escrow pending the adoption of the plan.[31] The employers refused this on the grounds that it might easily jeopardize the acceptance of the total package by the membership, and the proposal was dropped.

There are a variety of other fringe benefits, common to many collective agreements. Jury-duty pay, make-up pay during National Guard duty, bereavement pay (time off when there is a death in the family), call-in pay, and penalty pay on short-notice layoffs are the benefits which are most common in the lithographic industry. Paid sick leave is not provided in any of the contracts the writer has seen, nor is there any provision for excused time off. At one time, during the flurry of guaranteed annual wage negotiations in steel and autos, the ALA took up the issue in convention, but it has never become a factor in negotiations.[32]

The fringe benefits enjoyed by organized lithographers are above average but not unique; there are other work groups in both organized and unorganized plants whose benefits equal or exceed them. The more interesting aspect of benefits is the fundamental role of the union in administering them, chiefly because of the economies of scale which apply to the administration of such programs. A small employer in the industry can rarely risk the incurrence of such costs without the assurance that his competitors will also incur them, and it is significant that the greatest number of private pension plans is in the captive plants, where coverage of lithographers is a part of total plant coverage. Employer associations can be effective in the administration of health and welfare plans, but commonly lack the authority which is required for the central administration of benefits. The union is clearly performing a function normally considered a managerial prerogative, not only because it wants to, but because there is no managerial unit which can perform it as effectively.

Although one may expect that profit maximization will not be the goal of the union in administering such plans (for tax reasons if for no other), economic maximization of a sort is relevant. Commending the 1959 delegates for keeping their pensions solely union-administered, Benjamin Robinson said:

To me this is of extra significance in the days ahead, because of the constantly increasing anti-union atmosphere. Your problem now is to not only originally sell a man on union membership, but frankly, you have got to tie him to your union. You tie him to your union by doing a job for him. You tie him to your union also by his taking serious losses if he leaves the union. . .

It was our belief, and I think we are correct, that the longer the pension fund is in existence and a man has a great financial stake in it, all hell will have to burst before he will run away from that.[33]

The explanation highlights the split which can develop between organization goals and the individual goals of organization members. It also suggests that the acceptance of managerial functions may add some unexpected twists to the practice of trade unionism.

Chapter XI

Results of Bargaining:
Distribution of Power

Fundamentally, collective bargaining is the resolution of two issues: money and management. In the preceding chapter the money issue claimed attention, although it was impossible to discuss the question of fringe benefits without introducing the management issue. Here the focus is directly upon decisions about the distribution of power, decisions about who will manage what. Inevitably this involves the question of managerial prerogatives. The conventional employer view is that employers must man the barricades to preserve these at any cost. A view common among more experienced management bargainers is that bargaining over principles (such as managerial prerogatives) leads to bitterness and strikes, whereas bargaining over applications (such as the need to keep a half-dozen key workers during a layoff, regardless of seniority, in order to maintain operating efficiency) leads to understanding and solutions. A still more sophisticated view, integrating the two others, is that management must preserve the right to initiate action, protecting the workforce from unjust acts by conceding to workers ample right to challenge and appeal.

Implicit in all these views is the assumption that management is determining the disposition of power. In lithography, and perhaps in other industries, the approach is misleading because the assumption is incorrect. The disposition of power will be determined, not by people alone but also by the context in which the power is exercised. These are not purely political struggles, nor are the actors mere pawns in contests whose outcomes are economically predetermined. As elsewhere in this study, it will be evident that an industrial relations system operates in a context of several parts. Neither organizations, markets, nor technology can by themselves provide a satisfactory explanation of the distribution of power.

Controlling the Supply of Labor: Part I

This and the following section are concerned with a single topic — controlling the supply of labor. The division is made because two types of control are used. The first, considered here, deals with restrictions on new entrants to the labor force. The next section takes up restrictions on the hours of work which may be taken from existing members of the labor force. If the supply of labor is viewed as being made up of workmen, rather than of labor hours, then the first type of control is a restriction on supply, and the second a creation of additional demand.

Hiring provisions. In common with all unions which attempt to control the supply of labor, the ALA seeks to channel all new employees through its offices. There have been times in its history when organized employer resistance has challenged the union as the employment office of the industry, and when membership in the union was a distinct disadvantage in securing employment.[1] But for most of its history and certainly today, the most important disburser of lithographic skills in most localities is the ALA local union. This fact is the primary explanation for the large number of lithographic craftsmen who continue in the union when they work in non-contract shops. The dues they pay are the price of continued access to the union's hiring facility.

Employers, insofar as the writer has been able to determine, object to the union's control of the labor supply only in their public pronouncements. In practice the complaints, and there are many, emphasize the poor quality of the employment service provided. For example, during an interview, a Chicago employer pointed out that whenever possible, all craftsmen were appointed from among apprentices trained within the shop, and explained that "the guys the union sent were terrible." When this was checked with the Local 4 officer in charge of placement work, he pointed out that in Local 4 and probably in any local there are floaters, members who from incompetence, drunkenness, personality problems, or something else cannot hold a job. They are troublesome to employers and also to the union because they hold tightly to their union membership.[2] The union's grievance is that such people get in the union because management has apprenticed them without adequate screening, and has thus saddled the union with them after they have been fired for incompetence. No local official denied that on occasion "lemons" were sent out to employers.

Prior to Taft-Hartley the ALA had standard closed-shop hiring agreements in most of the major cities, and in common with a number of other unions, hesitated for some time before it decided to comply with the Taft-Hartley law. The importance of having the NLRB continue to throw up a protective shield for lithographic bargaining units soon settled the question, and hiring clauses were modified to conform with the prohibitions of the law.

Perhaps surprisingly, there is little uniformity among the hiring provisions in contracts of different cities. Here are two representative examples:

It is further agreed that the Employers will apply to the Local Union Office, when in need of help.[3]

Company agrees to advise the Union Office when in need of employees. Any person sent by the union office for a job, shall present his work card to the shop delegate after being interviewed by the Company and before starting work. Subject to the provisions of this Article, however, Company shall be at liberty to hire employees from any source it desires.[4]

New York adds a different twist. The operative clause is similar to those above, but in the back of the contract there is a provision that if Taft-Hartley is appropriately amended, a closed-shop hiring provision is automatically substituted.[5]

Hiring provisions usually apply only to the skilled classifications. Since apprentice vacancies are usually filled through promotion rather than hiring, there is a fairly clear breaking point. In some contracts, notably in those locals that have a fair proportion of general workers, there is a provision that the union will also be given first opportunity to fill the unskilled classifications. Probably a more common practice is the Boston hiring provision which applies to all classifications listed in the agreement; the general-worker group is not among those listed. Nevertheless, it was made clear in the meeting called for the explanation of the contract that the union wished to be notified of all openings, skilled or otherwise.[6] There was some question of what this would imply in specific situations, but none on the legitimacy of the union interpretation. Where the union is less firmly established, such a qualification may not apply. One of the Baltimore Can Company contracts has this narrowly defined provision:

The Company agrees to advise the Union office when in need of permanent skilled employees, provided that the Company shall not thereby assume any obligation to employ any person referred to the Company by the Union.[7]

It is unlikely that hiring practices would change materially in the absence of contract provisions. The hiring clause describes practice rather than establishes it, and its direct effect is probably limited to standardizing the procedures for accomplishing what would happen in any event, the union remaining as the central distribution point for lithographic skills in a single labor market, and the employers free to hire elsewhere only where there is a useful alternative source of employment.

It may be true that contract provisions slow down the development of alternative sources. Without them, trade schools might have developed more rapidly in the postwar period when the industry was expanding, yet the fact that plans moved slowly when the ALA and the industry associations were cooperating in the JLAC suggests that this is only a faint possibility. There seems little doubt that hiring provisions reflect the strength of the union but do little to add to it.

The union as personnel officer. Any local union must have some degree of veto power to be effective. It must be able to stop unfair firing, promotions based on favoritism, excessive work standards, and so forth. Jules Justin has identified this right as the "administrative right of protest and appeal."

As the exclusive representative of the employees in the bargaining unit, the union has the administrative right to protest and appeal every *action* that management's representatives initiate, which the union claims violates the contract. But a union does not have the right to protest and appeal management's right to act — management's right to initiate action — unless management's negotiators, by neglect or unwittingly, have given up that right by "prior consent," "mutual agreement" or "joint consultation" restrictions in the contract.[8]

Although Justin's view that management must retain the right to initiate action is sound, it overlooks the possibility that some managerial functions may be handled most effectively by the union. This affirmative union power must be distinguished from the more traditional veto power of unions since it rests on a different foundation.

The veto power stems from a realization by managers that they cannot function effectively without some degree of voluntary consent from the industrial workforce, and that the price of this consent is the right of protest and appeal. The acceptance of this condition by management has permitted the growth, in Professor Dunlop's terms, of a shared ideology, which in our society implies that free workers and free management can exist together in a free enterprise system.

Affirmative union power arises when there is no management unit to carry out a necessary management function. In the lithographic industry and quite possibly in several others, the managerial responsibilities undertaken by the union are similar to those of a personnel officer. Hiring is the clearest example, but by no means the only one. The local union president plays an active role in day-to-day contract interpretation, quite separate from his role in formal grievance procedure. In the Boston agreement, as in other group contracts, the essence of the terms covering working conditions is flexibility, bounded in some cases by rigid restrictions and in others by broader and more relaxed ones. It is a hopeless task to specify degrees of freedom in a written document, and inevitably the day-to-day decisions as to appropriate latitude in a given instance fall primarily on the local union president. Shop delegates call him; so do employers. The writer was privileged to hear one end of a telephone conversation in which the employer calling thought he had a contractual right to a transfer-layoff arrangement, an opinion which was not shared by President Arthur Willis of the Boston local. The employer proved stubborn, Arthur's words became less jovial, and the climax was reached with this phrase: "All right, Jerry, go on the contract, but just be careful what you do, because I wrote that clause."

A strong employers' association at the local level can reduce the power of the local union president by providing an alternative authoritative source of information, and in some cities such an association exists. But individual employers are likely to be more jealous of their freedom of action than shop delegates, and the grant of authority they give to their associations leaves the latter with far less authority over the management of an individual shop than the union president has over the shop delegate and workers in that shop. Unquestionably, the area grievance committee in Boston helps provide uniform interpretations, but no device in an industry of this sort is likely to relieve the local president of his responsibility as chief interpreter of contracts.

To take another example in the field of selection procedures, the Boston employers in the 1958 negotiations proposed that when the union sent a man for employment it also send "a complete and objective evaluation of the candidate's qualifications, his previous work history, and a record of his scores on appropriate Lithographic Technical Foundation Trade tests." [9] In the following meeting the employer proposal came up for discussion:

Union: We do not know we can take care of the LTF part of it. You are asking that a man must take these tests.
Employers: Yes. You say a man knows his job when you send him to us. You could [have him take them] in your office.
Union: The Employer can do it too.
Employers: Yes, but you send a man and claim he is capable of doing the work. It is your responsibility.
Union: No, it is yours. You hired him.

The union was adamant on refusing to administer the tests but agreed to give as complete an evaluation as possible. Several meetings later it counterproposed that when a man left an employer, the employer should give the union a statement of his experience and an opinion on his work. Said the employers:

We have no objection to your idea, but with the present high hourly rates that we have, we have got to make sure of what we are getting in the way of man-power. This is a responsibility of the Union and we want your members as a whole to recognize it.

They agreed that

The Employer will provide the Union office with a complete and objective evalua-tion of the work performance of each employee, in writing, such evaluation having first been reviewed by the Employer with the employee.

The union now had second thoughts:

On the matter of evaluation. For the sake of peace and harmony after the contract is entered into, let it stay as we have written it.

We have added to Section 1 the words "The Union will endeavor to supply the Employer with workmen of the capabilities required." We feel if we would put into the contract the word "evaluation" it would cause some trouble.

Employers: Suppose we take the Union's language and write you a letter of understanding as to what we want in an evaluation of a man insofar as you can give it to us. The scale going up as it is, we will have to have the best men possible.

Union: There is no argument on that.

Thus ended the proposal to write merit rating into the contract. The dis-cussion of this minor item in negotiations has been given in some detail because it illustrates the recognition by both parties that the union's role as employment officer made it a natural and in part a necessary participant in the process of determining relative worth of employees.

It has already been noted that the trade association performs some of the services of the personnel department for member firms. Its role is necessarily limited to those functions which are noncontroversial, or to put it another way, it can do those jobs of the personnel department in a large firm that the line foremen agree are useful and ought to be done. When the role is controversial, the trade association is ineffective. There is no management over both that can enforce the decision. The declining power of the national association in the industry has been described. What has happened is that it has been partially replaced by the union because the absence of a strong national association has pushed the collective relation-ship to a local level, where a single labor market has reinforced the power of the local. The role of the union as personnel manager for the industry is accordingly played at the local level, and in the controversial issues the union has the power to press its point of view strongly since it is the only actor in the relationship with an effective (though not complete) authority over the total workforce of the bargaining unit.

Apprentice provisions. Apprentice provisions provide controlled access to the journeyman classifications. The control is established by setting the number of journeymen who must be employed before a shop is entitled to make an apprentice, the length of apprenticeship, and specifying circumstances under which apprentices may be made.

The standard provision for the ratio of apprentices in most contracts is taken from the ALA constitution, suitably shortened to omit those provisions which are inapplicable in local situations:

> Sec. 1. For the first four journeymen in the Press and Process Art Departments and for the first five journeymen in each of the other departments except the Platemaking Department, Transfer Department and Poster Artist Department one (1) apprentice shall be allowed; and for ten journeymen two (2) apprentices, and for fifteen (15) journeymen three (3) apprentices, and one (1) additional apprentice for each five (5) additional journeymen in the department.
>
> In the Platemaking Department, Transfer Department and Poster Artist Department one (1) apprentice shall be allowed for the first seven (7) journeymen and one (1) additional apprentice for each seven (7) additional journeymen.[10]

The provision is interesting because it illustrates so clearly the effect of technology. Technically there is no such thing as a process art or poster artist department, and the "transfer department" clearly refers to transferrers, a part of the platemaking department. Although the union is committed to the proposition that journeymen should be broadly skilled and therefore apprenticed in a branch of the trade and not in a single classification, when setting ratios it wishes to pinpoint the classifications which must be protected. Prior to 1956, only the transferrers and the poster artists were on the 1 to 7 ratio; in 1956 the platemakers were added, a direct result of the rapid introduction of longrun plates. Many contracts, however, have not been altered to reflect this constitutional change.

Maintenance of ratios has long been an important plank in the ALA policy of controlling the labor supply. In 1904 the power of the union was such that the employers made a formal plea to have the constitutional ratios relaxed. After the 1906 strike, one of the points stressed by the employers' association to its member firms was the necessity of a full complement of apprentices at all times. The NAEL had established an apprentice ratio which varied by classifications, but went as low as 1 to 1 for transferrers and 1 to 2 for pressmen, in contrast with the standard 1 to 5 which had been enforced by the union prior to the strike.[11] As the union regained control in shops, it reintroduced the constitutional apprentice ratios, whereas the NAEL urged shops that it still controlled to keep the full ratio provided by the NAEL rules. The 1919 and 1920 national agreements avoided the question of apprentices entirely, and during the period 1920–1933 the establishment of ratios was less emphasized by the ALA than holding apprentices to their full term before they were allowed to act as journeymen.

During the NRA, apprentice ratios again became a subject for union-management negotiation and agreement, and from that time forward they have been contained in most contracts which the union has signed with individual employers or groups of employers. The acute shortage of skilled manpower in the war and immediate postwar period induced the international to allow a relaxation of ratios, and in the case of the artist department to dispense with them entirely for a one-year period. For this the international was censured severely by Local 1 on the floor of the 1946 convention, the local lamenting this "unheard of and extraordinary procedure" as striking at the very base of union security." [12]

In a memorandum the Chicago local attempted to show statistically that there was a predictable surplus of craftsmen in the immediate future, but it had to make some heroic assumptions to fit the facts to the prediction.[13] For example, all additional equipment coming in was seen as operating on one shift only, although at the time the memorandum was prepared nearly all shops were running two and some three shifts. One sentence summarizes the point of view which the memorandum was intended to support:

> Slow progress in the delivery of equipment . . . together with a possible slowing down of business resulting in nation-wide instability, causing unemployment, is a warning which should caution us not to train new craftsmen indiscriminately with reckless abandon by the score.

In other cities little effort was made to bring the ratio back into line, so that in 1947 there were still many more apprentices being trained than constitutional ratios permitted. Buffalo had a 1 to 1.4 ratio in the press department, and 1 to 1 in the art department. Cincinnati had an over-all ratio of 1 to 2.4, Detroit 1 to 1.4.[14] Racine held something of a record with fifty-seven journeymen and fifty apprentices. In the decade that followed, the situation from the union viewpoint has improved materially, but there are still locals with 1 to 2 or even 1 to 1 ratios in certain crafts.[15]

The period of apprenticeship must necessarily be controlled if the ratio is to have any restrictive value. The ALA constitution provides that this period shall be four years in all cases.[16] Although there is little question that four years is a legitimate period for apprenticeship in some of the skill classifications, it is clearly excessive for others, notably for the feeders. The purpose of a four-year apprenticeship in this case is not to provide an ample training period, but to keep the supply of feeders in balance with the supply of pressmen. Other methods, such as curtailing the ratio, would be more effective in achieving the same end, but the four-year apprenticeship period has a seventy-five-year history behind it and will not be changed without compelling reasons.

What primarily distinguishes an apprentice from a journeyman is that he is to work in several different classifications within the same department and is not to perform work in the absence of a journeyman. According to employers interviewed, this rule more than any other was the point at which relaxation was allowed. During debates in ALA conventions, allowing apprentices to do journeyman work so long as they were paid the journeyman rate has been openly supported. The ALA would seem to have placed its dependence on the ratio and an insistence that the formality of a four-year apprenticeship be completed (this in order to protect the restrictive value of the ratio). The Chicago contract actually does shorten the time for some feeders, though feeders on multicolor equipment still are required to put in a four-year apprentice period. The union has often permitted apprentices to move up rapidly in the wage scale when there was a genuine shortage of journeymen. For reasons of consistency the union would not wish to have such a clause appear in a contract, but when a union is secure, a contract can become a flexible instrument, and in an expanding labor market both union and employers would lose by enforcing this part of the apprentice provisions rigidly.

The final elements of apprentice control are restrictions on the making of apprentices. More broadly stated this implies alterations in the ratio, and earlier examples have been given of relaxation of ratios during the war period. Restrictions are aimed at avoiding unemployment in any single classification by cutting off replacements until the problem of unemployment is resolved. The strongest clause is one which assumes or explicitly states that the making of apprentices is at the discretion of the local, and then states the terms under which the local will permit management to make apprentices. Buffalo has such a clause; after specifying that the ratios shall not be more than those provided in the constitution, it then continues:

The union will, however, consent to the making of apprentices on the basis outlined above unless it is not warranted because of the unemployment status of the particular job classification in the industry, or other factors in the particular plant involved. If there is no mutual agreement the apprentice is not to be made and there shall be no arbitration.[17]

This language also appears in the New York contract, where it probably originated.

New York adds a new twist to the making of apprentices. The clauses are of sufficient interest to be quoted in full:[18]

If the Union and the Association agree in writing that there is sustained general unemployment in any job classification, then during the period of continuance of such joint agreement, (a) no apprentices shall be made in such classification, (b) qualified journeymen in such classification, from within the Employer's plant or new employees at the option of the Employer, may be made apprentices pursuant to Section 18 (b) in any other job classification at the journeyman rate of such unemployed job classification.

Such an apprentice who has been so made from the Employer's plant and has been replaced by another journeyman or who has been hired as new help may also work in his previous classification as long as the Employer does not reduce the number of employees in such previous classification which it had prior to the making of such apprentice and provided the amount of work of such apprentice in his previous classification does not justify regular additional employment.

Lack of agreement as to the status or continuance of sustained general unemployment shall not be subject to arbitration.

Coupled with this are several other clauses intended to provide training and retraining facilities, so that already well-developed mobility within classifications may be extended to mobility between classifications. One of the most significant provisions combines nine preparatory classifications in 1948 into a single one today, with provisions for forceful encouragement of members of the dying classifications to go to school and, in President Swayduck's phrase, "rehabilitate themselves." [19]

Other locals, notably Chicago, have also moved far in combining classifications. By so doing, a decline in the demand for one type of skill can be met with minimum dislocation and resistance from the affected workers. Fewer classifications, with more broadly trained craftsmen, also facilitate intershop mobility, which as already noted is a prime requisite if premiums above scale are to be secured for any significant portion of the union membership. Yet combining classifications has several advantages for employers also. Flexibility in the use of workers, a major advantage in any job shop operation, is materially increased. A clause in the New York contract, for example, specifically provides that "Employers shall have the right to have Platemakers strip."[20] (Numerous shop comedians have doubtless bewailed the fact that all platemakers are men.) Even in the absence of such specific provisions the employment of a broadly trained craftsman permits the employer to experiment in new product markets without hiring new men. In addition, worker mobility, which a well-trained workforce facilitates, favors the successful and growing employer at the expense of the stagnant or marginal ones.

Initially, the effect of combining classifications is almost certainly an increase in labor costs. But over the long-run the scale rate for the combined classification may rise more slowly than rates for some of the high-skill jobs which go into the combination, and the use of premiums for individual differences may well replace former scale differentials, with an accompanying improvement in the degree to which wage payments reflect value received from individual workers.

Local leadership in the ALA is outstanding in its ability to persuade the membership to agree to and cooperate with the introduction of innovations in the industry. For one thing, the structure of the union is such that so long as the industry is expanding, there is always the possibility of a member shifting from a declining to an expanding craft. In this the ALA is significantly different from the other printing unions. None of the other unions has a series of approximately equally skilled craft groups within its membership, and as a result these unions encounter far more resistance to the introduction of innovations. Because the ALA need fear this less, it can afford to be more flexible on apprentice provisions, and can use them instead to capitalize on the multicraft nature of the union. For the same reason, it is possible for the leadership to look with more favor on the longrun benefits of innovations. Because the union realizes that other unions are trying to control the expanding labor force, it has an interest both in having skilled men available to new employers who introduce the process, and in having these employers view the ALA as a union which recognizes employer problems. This interest, if followed exclusively, would lead to a policy in conflict with the apprentice restrictions. The interest is not followed exclusively, but neither is it ignored.

In recent years there have been particular efforts to look at the apprentice program from a different viewpoint. The nature of the attempts to bargain with employers on a national apprenticeship policy, for example, has indicated that the union has not been interested in setting nationwide restrictions as much as in setting national standards of training adequacy. Among the locals, New York and Chicago are notable for their present efforts to keep the labor force in balance but not to hold back the expansion of the industry through a shortage of trained manpower. Presidents of these locals today can sympathize with some of the problems of the NAEL secretary in the period following the 1906-1907 strike. Like him, they want neither too many nor too few apprentices trained. As early as 1946, in the same convention in which Local 1 was lamenting the "unheard of and extraordinary procedure" of allowing unlimited artists apprentices, in the same year that Chicago was cautioning leaders "not to train new craftsmen indiscriminately with reckless abandon by the score," Benjamin Robinson made the following statement to the ALA convention delegates:

> The theory of a scarcity of trained workers to protect present jobs and to force higher wages is a short-range policy to the advantage of relatively few workers or for particular classifications of workers. It would lead inevitably to the end of control of lithographic workers by the Amalgamated Lithographers.
> The organization's approach must be a flexible approach, noting that rules made 15 or 20 years ago do not necessarily apply to the conditions of today . . .
> Not all of today's classifications of work require four years of training. *New developments in black and white production have created jobs for people which do not require our traditional four years of apprenticeship. You are going to have to do something about this.* Black and white photolithography is an integral part of this industry just as much as four-color reproductions of fine art subjects. The failure of the union to keep pace with the facts of life in this industry will force equipment manufacturers and employers to train men themselves, to train men in schools beyond the control of the union and make the way easier for Printing Unions to break into the industry.[21] [Italics the author's.]

This was the kind of thinking that had led to the formation of the Joint Lithographic Advisory Council. Another kind of thinking led to the 1946 negotiations in New York, which as already noted heralded the end of effective union-management cooperation in the industry. The attitude typified by Robinson's speech in 1946 probably prevails in the international leadership today, but the long interval between has seen a rapid expansion in the lithographic process, much of which has not come under the control of the ALA.

In many respects the union has not yet made up its mind which policy to follow. The leadership, notably Ed Swayduck, recognizes the impact innovations will have. The clear thinking on the problem that is illustrated in recent conventions and in articles in the journal is rare among unions. But the abysmal mismanagement of the offer to cooperate with employers on harnessing these technological developments (the "million dollar offer") comes not only from the Local 1 president's desire to have the *New York Times, Wall Street Journal,* and management groups grant him the accolade of industrial statesman; it also stems from the union's inability to realize that its long-term security *as an independent body* requires a firm working relationship with lithographic managements of the kind that Kennedy was able to fashion during his ten-year reign. There is always the prospect, of course, that the ALA will choose not to continue as an independent body but will join with the ITU or some other union, and thus reduce the need for employer cooperation to achieve long-term security. In any event, cooperation will not come through a union version of Boulwareism. It will come only when the ALA realizes that it may have to give up some of its traditional definitions of "good" working conditions, if the result of innovations is to be a further growth of the lithographic industry and not a further scattering of the lithographic process.

A union of 36,000 members cannot indefinitely afford the costs of following the process into already organized industries. The least of these costs is the substantial legal fees the ALA pays each year; more significant are the hidden costs, for example, organizing time spent on court and NLRB proceedings, consulting with attorneys, learning the law, and most serious, learning to give first priority to the legal rather than the human challenge of organizing. When "timely filing" becomes the test of an organizer's excellence, the novice wisely spends his nights studying cases rather than calling on prospects.

The ALA has not now, nor has it ever had, a chance to eliminate completely the spread of the process to other industries. But a policy of testing apprentice ratios, training periods, manning requirements, and similar constitutional provisions against current rather than past conditions, and of selling the permissible concessions to employers for joint research and extensive training, might well have slowed the scattering of the lithographic process.

Seniority provisions. The union's attitude toward seniority reflects the power of the local union. Few contracts have provision for action based on seniority except in the case of semipermanent layoffs or a reduced workweek. Even in the case of choosing apprentices there is seldom a strong emphasis on seniority. The Boston contract says nothing on the subject; New York's provides that "the selection of employees for apprenticeship shall be based on competency and length of service"; Chicago's, that the seniority rule shall be followed "as closely as possible," but with no reference to the reasons that would make rigid adherence impossible.[22]

In the 1955 convention a resolution favorable to seniority was opposed by Local 1 President Swayduck with the following argument:

> Seniority, as used in some shops, is a boss's weapon . . . I mean by that the following: when a man wants a wage increase, a premium increase, the management knows that that man has built up seniority, and should he leave his shop, the next shop he goes into, he has to start building a new feeling of security. He's got a weapon. That man will stay put and forego a request for wage increases.
>
> We enjoy a large premium pay in my area, because we have always made that very clear to the members. Seniority means nothing. When you leave this job, there is another job.
>
> We propound the theory that there should be nothing to bind a man to any particular job by virtue of having to wait for a certain amount of years before he is secure in that job.[23]

This statement was received with applause, and a number of other speakers supported Swayduck. Strong opposition came from the Austin, Texas, local, whose delegate pointed out that in states with right-to-work laws, seniority gave some protection to union members against discrimination. Other references were made to the usefulness of seniority in metal decorating plants. The relationship was clear: where mobility is low to begin with and/ or the local is weak, protective seniority provisions will not be considered a "boss's weapon."

Issues of job control. There are relatively few contract provisions which restrict output in the lithographic industry. In a sense the complement clauses should be viewed in this light, but their primary purpose is to further uniform wage costs throughout the country. A number of contracts write in the constitutional restrictions on piecework, on running two presses or other pieces of equipment, and on team work, which is in essence de-skilling an operation through job breakdown. All of these provisions have more historical than current interest. The constitutional clauses go back at least to amalgamation, and some to the 1890's. Their current interest may be summarized briefly.

Piecework is a problem of the trade shop. It offered a particularly appropriate incentive for engraving work, but when that operation began its decline with the introduction of photographic platemaking, it ceased to be as important an issue. Currently there are few jobs which lend themselves readily to piecework, even in trade shops. The importance of having the union label insures that few trade shop employers will risk the rupture in union relations that would almost certainly follow if piecework were to be introduced.

Running two machines is currently no issue at all, though perhaps it should be in connection with offset duplicators. The union has fought the two-machine issue successfully on all types of equipment over the past seventy-five years, and is openly contemptuous of the IPP&AU's willingness to let its members run more than one press. But fighting for job control on an operation which can move out of the industry only with difficulty has a rationale that is irrelevant when the operation has great mobility. The insistence of the ALA on treating Multilith operators as pressmen has probably speeded the flight of this type of equipment into combination shops, open shops, and private plants.

Teamwork is opposed by the ALA because it destroys craft skills, and, therefore, the basic source of the union's power. There are grounds for believing that in some types of firms, notably those which specialize in the production of a single product or a closely related group of products, task specialization would lead to more efficient production routines. The ALA fights this but not because it wishes to make work for its membership. There is a genuine conviction that the thoroughly qualified craftsman is a more productive and more valuable workman for the industry as a whole, and much experience to prove that he is easier to place if the need arises. Teamwork, because it destroys craft skills and replaces them with narrow specialties, does destroy mobility, and without mobility, workers are prone to resist technological change. On balance, it is unfair to classify the union opposition to teamwork as featherbedding, although in some shops it undoubtedly leads to an inefficient distribution of work.

Controlling the Supply of Labor: Part II

The thirty-five-hour week. There is no doubt that the motivating force behind the ALA's drive for shorter hours was the fear of unemployment. It has been noted in connection with the apprentice clauses that this fear led at times to an almost irrational attitude on the part of some union leaders in the postwar period, a fear which does not appear to have been shared by many union members. The transcript of the 1947 negotiating meeting is strewn with references to the lack of member interest in shorter hours, and in spite of the fact that current union mythology requires that the demand for the thirty-five-hour week be considered as welling up from the rank and

file, even in Local 1 there is some question whether the shorter-hours demand gained more impetus from the floor or the rostrum.

The fear of unemployment in the leadership was well founded. The
period from 1927 to 1939 had given the union long experience with unemployment, and for much of that time it was the dominating element in
union calculations of its bargaining strength. Figures on unemployment and
short-time for a portion of that period are given in Table 17.

Table 17.   Unemployment in the ALA, 1932–1939.

Year	Journeymen members	Per cent journeymen unemployed	Per cent journeymen part-time
1932	4,845	24.9	47.3
1933	5,180	17.7	24.5
1934	5,720	14.3	20.3
1935	5,733	11.7	18.5
1936	5,943	7.9	9.9
1937	5,884	9.4	19.7
1938	5,870	13.0	24.6
1939 (June)	5,710	10.6	17.7

Source: Report of President Kennedy, ALA *Proceedings* (Atlantic City, 1939), officers reports, p. 7.

Efforts had been made to reduce the workweek to thirty hours during the
depression, and when the first opportunity came to bargain without governmental controls, Local 1 of the ALA went into negotiations with a
demand for a seven-hour day and a thirty-five-hour week.[24]

The settlement was for 36¼ hours, effective July 1, 1946, and was followed by hours settlements in other parts of the country which established
either 37½ or 36¼ hours as the workweek for 80 per cent of the ALA membership in less than a year.[25] Hours remained at this level until 1953, when
Chicago broke the hours to thirty-five, effective January 1, 1955.[26] In the
same contract the third-shift hours were established at thirty, the second
shift remaining at thirty-five. Following this settlement there was a similar
rapid progression in hours reductions across the country, so that by mid-
1957 only scattered locals, mainly in the south (and in Canada), were still
on 37½ or 36¼.[27] The thirty-hour third shift spread less rapidly, being in
force only in large locals, with about fifty per cent of total union membership.

By 1957 the ALA had clearly moved ahead of the other printing trades in the goal of hours reduction. Only the photo-engravers came close, with 52 per cent of the membership on thirty-five hours, 36 per cent on 37½, and the balance in between.[28] Hours reduction in the lithographic industry had traditionally lagged behind that of the rest of the printing industry as Table 18 shows. Beginning in the postwar period, leadership was reversed, and lithographers led the letterpress unions.

Table 18. Prevailing hours in the unionized printing industry.

Year	ALA	IPP&AU
1895	54–59	54
1907	54	48
1911	48	48
1921	48	44
1927	46–48	44
1933	40	40
1947	36¼–37½	37½
1955	35–36¼	37½
1957	35	36¼–37½

Sources: IPP&AU prevailing hours are based on hours for cylinder pressmen in book and job printing, as shown in BLS, *Union Wages and Hours: Printing Industry*, Bulletin no. 1228 (May 1958), pp. 15, 17. ALA prevailing hours (LIP&BA before 1915) are drawn from Hoagland's study for the years prior to 1915; from material presented in the historical chapters of this study for later years.

From July 1946 to July 1957, there was only a 4 per cent reduction in scheduled weekly hours in the printing trades generally (excluding the ALA), the largest single one being a 7 per cent reduction for the electrotypers. For the ALA the reduction was slightly in excess of 10 per cent, and closer to 12, if the Canadian membership is excluded.[29]

There have been some assertions that the rapid descent to thirty-five hours has penalized the lithographic industry and caused some work to move from lithography to letterpress. It must be remembered that an hours reduction in the printing trades always has two elements: a reduction in the workweek and an increase in hourly rates to maintain weekly earnings. Accompanying wage increases, if any, are always in addition to the amount necessary to maintain weekly earnings. A number of industry spokesmen have asserted that work shifted to letterpress plants when Chicago lithographers dropped to thirty-five hours in 1955, but evidence that such a shift did in fact occur is difficult to find, and to the writer's knowledge no one has succeeded in doing so. Vice-President Rohan of the IPP&AU when questioned on the subject said he had heard there was a slight increase in work for letterpress shops, but clearly did not attach much significance to it.[30] The competition that worries lithographic firms is still from other lithographic firms, whose wage bill and working conditions differ from their own either because they are unorganized, organized by the IPP&AU, or in a different local jurisdiction. Yet we have seen that this competition, particularly from the open and IPP&AU shops, is important and growing. When in 1959 San Francisco submitted a resolution to introduce the thirty-hour workweek to the industry, a delegate from Kansas City (in which IPP&AU offset membership is substantial) opposed it with the following statement:

The IPP&AU still works a 37½-hour week. To reduce our work week to 30 hours generally would sometimes put the employers on a non-competitive basis with the lithographic pressmen who belong to the IPP&AU. Therefore I believe we should leave this matter to the discretion of our International Councillors and our Vice-Presidents.[31]

Local 1 in New York also opposed the resolution (for additional reasons to be discussed later in this chapter), but to avoid an open rejection, the proposal was accorded the "decent burial" of a referral to the international council.

Night shifts. In keeping with the emphasis on reducing straight-time hours, the ALA has in a number of locals reduced the night-shift hours (with no change in the weekly base rate) in addition to negotiating night-shift premiums. New York, Chicago, St. Louis, and the Pacific coast locals, which together have 50 per cent of the membership, have a thirty-hour third shift; the Pacific coast locals also have a 31¼ hour second shift. The remainder of the locals have with few exceptions thirty-five hours for all three shifts. Substantially all locals have night-shift premiums, typically of 10 per cent for the second shift and 15 per cent for the third.

In the opinion of several Boston employers with whom the subject has been discussed, a six-hour third shift would raise labor costs per unit of output higher than if the same output were produced on three hours overtime at the end of the first and second shifts, even if the third overtime hour on each shift was at double time. This is particularly true if through overtime hours it is possible to hand over a running press to the incoming shift and avoid the nonproductive time associated with protecting the plate on a standing press and preparing the plate. No one to the writer's knowledge collects overtime hours by shifts, or even segregates night-shift data which are collected. No estimate is therefore possible of the increased hour costs associated with the thirty-hour week.

Overtime. There are two aspects to overtime provisions — the amount of premium and the circumstances under which overtime may be worked. There are the normal variations among cities, but the typical provision is for one and one-half time for the first two or three hours beyond the scheduled shift and double time thereafter. In addition, a person working more than one and one-half hours overtime is granted a half-hour supper period, paid for at the overtime rate. Time on Saturday, Sunday, and holidays is at double time; and some contracts provide for triple time for work beyond seven hours on a Sunday or holiday. All this is in addition, of course, to holiday pay. Inasmuch as the annual average of weekly hours worked has remained well above the scheduled hours in the industry, the rate of overtime payment becomes a significant determinant of take-home.

In only one month (April 1949) during the more than a decade covered by the list [32] below have the average weekly hours in the lithographic industry fallen below thirty-eight.

Year	Hours[33]	Year	Hours
1947	41.4	1953	40.6
1948	39.5	1954	40.0
1949	39.3	1955	40.2
1950	40.0	1956	40.0
1951	40.1	1957	39.4
1952	40.2	1958	38.9
		1959	39.6

It has already been noted that restricting overtime is a union device for putting pressure on employers during negotiations. The contracts reviewed provide that "overtime shall be permitted when necessary," in order to specify that the union's power should not be used to withhold labor in the normal course of business or to protect employees who might do so. The Boston contract spells out this feature of overtime provisions more clearly than do other contracts.[34]

The employers' position was explained by Kenneth Scheid, of Forbes, a member of the employer negotiating committee, during the meeting called to explain the contract to employers and shop delegates:

Let me state what I believe is the understanding. There is a violation of this contract and it is a violation of Article XIV, Section 1, and Article XVII which is the no-strike clause, whether it be that is, [when there is] a mass refusal of a group of employees to work overtime. This does not extend any liability to the union. It is a risk, let us say, that the employees take. The union has certain steps that they take in the case of a contract violation of this nature. It is . . . [one of the purposes of this meeting] to avoid getting to that situation. Of course, the individual employee is perfectly free to decline overtime but there is one question, what about the group getting together as a whole and for some purpose? It usually involves a grievance of one type or another, stating a refusal to work overtime. That is the type of thing we are attempting to eliminate through a better contract and area grievance procedure and so forth. It has to be mentioned. It is in the the contract here.[35]

Other restrictions on overtime apply when there are workers laid off, and, though it is provided that overtime may be worked under such circumstances in an emergency, the burden of proof is on the company to show that an emergency exists:

However, in an emergency to service a customer an employee may work overtime, but a laid-off employee shall be entitled to pay for these hours if by grievance procedure it is determined that the work could have been performed by him on a regular shift. During the processing of the grievance no further overtime shall be allowed in the classification.[36]

Holidays and vacations. Holidays and vacations, like the thirty-five-hour week, have the effect of reducing the hours available from the existing workforce. There is less uniformity in these provisions than in either hours or overtime. Holidays average around seven days per year, Boston leading with eleven and New York with ten. In vacations the norm is three weeks after three years, though this is progressing to a standard three weeks after one year. A number of locals have secured two weeks after one year and/or three weeks after two years, but as of mid-1958 only New York had the three weeks-one year provision.

The international policy, which the majority of locals have not been able to implement, is to have no vacation accruals for length of service in a firm, since such a benefit tends to inhibit mobility between shops. In New York, vacation accrues from the beginning of employment, and payment for vacation credits earned is made if the employee leaves before the vacation period. The shop to which he shifts is obligated to give him vacation time during the vacation period, so that in effect the employer group is treated as a single employer for purposes of accruing vacation credits. Syracuse also has a similar arrangement, but, for the present at least, such contract provisions are rare. They are quite naturally resisted by the employers, and are not the sort of thing that stirs up member enthusiasm. There is a kind of commonsense justice in giving long-service employees something extra that appeals to the typical member of a union bargaining committee, the men on such a committee not usually being junior employees. To a long-service member, four weeks' vacation after twenty-five years' service with a company is a valued concession, whereas to the international it is a step in the wrong direction.

Attitude of membership toward hours control. The demand for shorter hours, starting in 1946, clearly arose from the carry-over fear of unemployment. By 1953, however, when Chicago signed the agreement for thirty-five hours, it was much harder to make this danger of unemployment meaningful. Today, nearly twenty years have passed since unemployment has been a serious problem for the union. The average age of ALA members is less than forty, and the average time in the industry is well under twenty years. It is understandable that arguments based on the tribulations of the 1930's carry little weight with the membership. In point of fact, what sells hours control devices today is not fraternal concern for the unemployed, but a desire to increase the amount of overtime hours that can be worked. ALA officialdom deplores overtime, particularly if there is even a trivial percentage of unemployment in the local. This is understandable since unemployed members are highly visible to the leadership. In a typical working day in a local of a thousand members, with ten of them out of work, the union officer will see and hear from more unemployed than employed members. Thus there is continuing divergence between the leadership and the membership over the purpose of the hours control devices. For example, in the Boston negotiations issues arose several times in which the official union position that overtime was an evil came into conflict with the known desires of the membership.

One such case concerned working through the lunch period. The old clause provided that "no employee shall be compelled to work more than five hours without being permitted to have a lunch period." The following discussion took place on the union's proposed change:

Employers: What about the "may not be *permitted* to work more than 5 hours without a lunch period." The word "compelled" has been used in the present contract.
Union: Yes. Before it said compelled. We are now saying "permitted," because some men work through their lunch period.
Employers: What you want to do is to eliminate 7 hours straight [time] work.
Union: That is what we mean though it doesn't say it. If the company wants a man to work thru his lunch hour then there should be compensation paid for the time worked.[37]

A clearer instance of controlling the amount of productive time available from employees is a rather common provision concerning vacations:

> The purpose of a paid vacation is to give the employee the opportunity of rest and relaxation and he shall not take employment elsewhere *in his craft* during his vacation period.
>
> No employee shall receive pay instead of vacation time except as may be mutually agreed between the Employer, the employee and the Union. [Italics the author's.]

The Boston contract also includes a general prohibition aimed directly at the members:

> No regular employee of any shop under contract with the Union shall work at the trade for any other lithographic employer without permission from the Union and his regular employer, and any employee working elsewhere with permission shall be paid regular overtime rates of pay.

"Moonlighting," it should be noted, is prohibited only in the trade. In this case, however, the issue was first raised by the employers, and the provision inserted to stop key employees from helping competitors. The union agreed, since it also wished to ensure that employees would not use their free time to defeat the purpose of the hours control devices. In spite of such restrictions, the problem remains serious, particularly when members will accept and sometimes demand overtime even when unemployment exists in the local. Ed Swayduck dwelt heavily on this situation in his remarks opposing the thirty-hour-week resolution during the 1959 convention:

> Getting the 30-hour week to guarantee the Brothers another five hours' overtime is not the intention of a trade union. You will still have unemployment.
>
> Until we can get our Brothers educated, and we should do that starting right now, that first we have got to harness the overtime hounds, teach them that one job is all they should have, not a job and a quarter or a job and a half, then talking about a shorter work week will make sense as we move along over the years . . .
>
> How can a man sleep nights when there is unemployment in his Local, when he doesn't say to his boss, "Look, I am not working overtime when there are men unemployed." Instead it is, "What do you have?" — a man pressuring the boss, "If I don't get overtime I am going to quit," and if the man is a good key man he is given the overtime at the expense of another brother.
>
> So let's start a campaign in our own organization. If you want the shorter work week, the 30-hour work week, let's first work 35 hours a week, and work on one job at a time.[38]

Control of Job Opportunity

The control of job opportunity is an issue which transcends collective bargaining, because it is only in part an issue between the union and the employers. A more important arena is the NLRB because the ALA has chosen to protect its members' rights to jobs by leaning heavily on the board's power to determine appropriate bargaining units. Favorable board determinations, however, are of no value unless the ALA can organize the workers in them. Thus organizing activity is the most important method of securing job rights. The ALA has ample power to protect the job opportunities of its membership from any employer action in the great majority of firms in which it has bargaining rights. Accordingly, union shop provisions and the check-off are seldom an issue. (A chapter of exceptions could be written, but it would refer only to southern locals in the current period. Historically, however, these have been issues in major ones as well.) The challenge comes from firms which the ALA has not yet organized, and unions with which it has no formal relations, or very tenuous ones.

The basic problems derive from the effectiveness of competition: the ease with which customers of the lithographic industry may become customers of another industry, with which an operation may move out of the lithographic industry into another industry, with which other firms may move into the lithographic industry, or semiskilled workers on inexpensive machinery may produce some lithographed products. So long as the ALA controls the supply of lithographic skills, it can cope with the mobility of the product, the mobility of the operations, and the ease of entry into the industry. But the easy-to-operate small offset presses and simplified plate-making equipment which were introduced during the 1930's required little skill, and therefore did not need craftsmen to operate them. The workers on these presses became members of other unions and in time increased their skills as firms with the Multilith added a 17″ x 22″, took the leap to two-color, and finally became full-fledged color lithographic plants. This process is by no means complete, for the ALA still bargains for the great majority of skilled lithographers in the industry. But it is a continuing process, and one which has worried leaders in the ALA since 1944:

> Unorganized *existing* shops do not constitute a serious problem today.
> Tomorrow, these same shops will be important, and will be joined by many of the new shops to be set up. As unorganized shops, they will breed trouble from at least three directions: competing with and threatening our standard working conditions; adding strength to increased open shop activities; and

leaving fertile territory for organizing activities of other unions. Today these troubles exist only in minor degree; tomorrow they will be more serious.[39]

Declining control of the labor supply has meant in effect that the union could no longer wait for employers to come to it and ask for help; it had to go out and fight for the right to represent men who produced in competition with organized lithographers, if the job opportunities of the membership were to be preserved. It had to organize lithographers in the Sutherland Paper Company if the jobs of members lithographing cartons were to be preserved, to keep a firm hold on the expanding trade shops if the jobs of preparatory craftsmen were to be preserved, and to organize lithographers in combination shops if the jobs of members in the smaller commercial lithographic shops were to be preserved.

To some extent the union has found it possible to use existing collective relationships to assist it. Because the ALA does have a large number of combination shops under contract, it has sought to protect its position in these firms by securing a contractual right to any future employment on "new or improved machines or processes for lithographic production work."[40] A related clause discourages the company from removing any work from under ALA jurisdiction and giving it to another union. These clauses have come under no serious attack by the employers, although some have resisted them for the quite practical reason that they arouse antagonism in other printing unions represented in the shop.

Another group of clauses has been challenged more vigorously. In order to strengthen its position with respect to trade shops, the international has urged locals to seek employer agreement that the contract has been "negotiated on the assumption that all lithographic production work will be done under approved union wages and conditions." The purpose here is to ensure that any trade shops used by the employer will be under contract with the ALA. The "struck work" clause is similar and is intended to discourage any employer from handling work for another employer who is being struck by the ALA. Closely related is the "chain shop" clause, freeing members from observance of the no-strike clause if another plant of a mulitplant firm is being struck by the ALA.

To the unpracticed eye most of these clauses appear to compel employer acquiescence to a secondary boycott, and therefore to an unfair labor practice under 8(b)(4) of Taft-Hartley. Some highly practiced legal eyes have also seen it this way, among them Earle K. Shawe of Baltimore and former LNA Counsel Gerald Reilly. With the benefit of their and others' advice a number of employers have successfully resisted putting the clauses in their contracts. Yet because employers tend to weigh the costs of fighting for a principle as carefully as they weigh the costs of a wage concession, the legal question did not reach the courtroom or even the NLRB for several years; it was by no means definite that the clauses were illegal. They do not say that union members will refuse to handle "hot cargo," only that if the employer fails to comply with the terms of the contract, for instance by asking employees to handle struck work, "the Union in its discretion, by notice in writing, may re-open the contract for negotiations as to the whole or any part thereof." As Robinson said to the Boston employers:

We negotiate this contract on certain assumptions that all work will be done under Union wages and conditions. We do not say we will not handle this type of work, but if you do [ask us to], then we have certain rights in the matter. This paragraph has been drawn with care on our part. Not only to reach our objectives by preserving our assumptions, but also to be able to sustain our position in any court.

More or less by accident, the first major test came in San Francisco, when the ALA local membership rejected a rather substantial package including the language, and struck for more money.[41] With all issues re-opened, the employers sought aid from the NLRB regional director, arguing successfully that all the clauses were illegal under the new and stronger secondary boycott provisions of the Landrum-Griffin Act. The battle was joined in the hearings before the United States District Court to determine whether any strike to secure these clauses should be temporarily enjoined. The outcome was assent to a temporary injunction (the actual injunction was never issued) against striking for the "trade shop" and "refusal to handle" clauses, and also against the "struck work," "chain shop" and "termination" clauses unless, as the ALA urged, the union was able to secure the court's agreement that appropriate clarification and revision had freed the last three from any taint of illegality. Since, under the law, the court needed only to be shown that the NLRB regional office had "reasonable cause" to consider the clauses illegal in order to grant the relief requested, it was not a sweeping employer victory.[42]

The union came off with a compromise since the final contract settlement in San Francisco included the contested clauses wtih the provision that they would be inoperative until proved legal.

Why did the San Francisco employers give even this much? Probably the most important reason lay in lack of employer unity. Employers of the 1200 ALA members on strike were carrying the ball for employers of the 1000

ALA members still working, employers who agreed to abide by any settlement reached. If work shifted from struck firms to operating firms, as it probably did, the parallel to scabbing becomes very clear indeed. In addition, employers were not deeply concerned with the language issue. They had already conceded it in the settlement rejected by the membership, and had adopted no basic change in position since that time. Employers were more interested in getting back into production under some reasonable economic settlement.

The San Francisco case has started the legal chain of events by which the validity of the clauses will be determined, and the ALA may in time lose its right to use them. Yet one may as easily overemphasize the ALA's potential long-term losses as its short-term gains. The attention centered on "the language" by employers and union people alike derives from a fascination with the sweeping implications of such language, if some set of specific occurrences takes place. But the "if" is a big one, and the language remains important only as insurance for the union, while it does little directly to advance job rights. Job rights are secured primarily by organizing and secondarily by limiting, through the NLRB, the number of workers who must be organized into lithographic units. Facilitating this through contract provisions with existing employers is less important than either getting the voters or defining the unit, and is less important an issue between the parties than other issues which can be resolved only through collective bargaining.

Chapter XII

The System: Present and Future

The industrial relations system in the lithographic industry has several features which distinguish it from other systems.

Prior to the 1930's the determination of working rules was a national question, but during that decade changes in the context of the system brought bargaining to a city-wide level. The basic change was an expansion in the importance of the local market for lithographed goods, which benefited lithographic employers but weakened their bargaining power with respect to the ALA for the local product market was roughly contiguous with the local labor market, and the ALA, being a craft union, found its basic source of power in control of the labor market. To a degree this control perpetuated itself; by controlling the existing labor force the employers came to the union for workers, and, because they did, the union was the only effective employment service. This prevented employers from training a surplus of workers by making an excessive number of apprentices. Having a share in the control of the apprentice program also permitted the union to take full advantage of the multicraft nature of the union because when the demand for one craft skill was declining more rapidly than the rate of attrition, journeymen in such a craft could be retrained in other skills. Thus there was small loss in supporting technological change, and much potential gain, since the market was expanding rapidly, and the unions' bargaining power could be used to insure an immediate sharing of the benefits.

This combination of circumstances has made it possible for the union to take a less jaundiced view of technological change than other unions in the printing trades, and featherbedding is both historically and currently less common in lithography than in the other branches of printing.

The same developments which have caused bargaining to move to a local level and encourage the union to support technological change have made firms using the two printing processes, lithography and letterpress, active competitors. The operation of the buyer-supplier bond has protected letterpress firms from losing their customers until they have added lithographic equipment, and these combination plants, often organized by other unions, add to the pressure for local uniformity in labor costs. The growing number of small firms has meant a growing number of unorganized firms. These developments make it increasingly difficult for any union to have effective control in the labor market.

The continued importance of the national market and the power of the ALA provide strong pressures for national uniformity of wage costs, as has been evident by the rapid spread of the thirty-five-hour week and health insurance and pension programs. National uniformity of wage costs is a central feature of the international union's wage policy since the rapid gains in working conditions in the postwar period have been made possible primarily through advances in the processes which affect these markets. The union is anxious to protect these gains and secure further ones, but to do this it must keep interregional wage costs roughly uniform among plants that compete in this national market. The obvious conflict between bargaining policies based on national and on local considerations is lessened because the same firms and the same skill classifications do not serve both markets. But the conflict is not avoided. There is in the first place a substantial overlap among markets served by individual firms, and added difficulty in that certain working conditions, notably hours, must be uniform for all firms and classifications covered by the local-area contract. For this reason the method of local group bargaining is no longer perfectly stable in determining working conditions in the industry.

The changed organization of industry and certain elements of the product market structure have eliminated the possibility of employer coordination of bargaining strategy at the national level, and have reduced it at the local level. When lithographic product markets were served primarily by lithographic firms, it was possible to resist union demands through taking a strike if necessary. The great increase in the number of firms and the spread of the process to other firms have so reduced the homogeneity of employers that such cooperation is no longer possible. Three national employer associations reflect in part the divisions among color lithographers, commercial lithographers, and combination plants, but these distinctions are of purely historical significance as firms expand into new markets, and as the associations, like the union, expand their jurisdiction to cover "all lithographers wherever they may be." This lack of unity is accentuated in localities where there are firms that draw on the same labor market but face widely different product markets. The trade shop owner, the small commercial lithographer with a secure group of house accounts, and the large color and publications lithographers serving their customers under a supply contract have little but the fear of whipsawing by the union to hold them together. As a result of this inability to offer effective resistance to union demands, the union has been able to secure immediate benefits from the gains in productivity associated with recent innovations.

But in the longrun the power of any union depends on its control over the supply of labor. In spite of its gains in the war and postwar period, the ALA's power in the labor market has visibly declined. Part of this decline is a direct result of its gains in wages, hours, and working conditions. When two unions compete, the preference of the employer may be decisive in determining the choice of the workers between them. The spread of the process outside the industry has brought the ALA into competition with many other unions in its efforts to retain control over all labor that produces lithographed products. In practice the international and its staff of organizers have concentrated on protecting the national product market, and have been less active in protecting the local market, where the union is a relative newcomer. Local organizing has been left to the locals, and its effectiveness has quite naturally varied widely. A union which has only thirty-six thousand members could do little else, but, as a result, the IPP&AU and other unions have rapidly increased the number of their members working on offset. The introduction of photo-typesetting has added a second major competitor, the ITU, which is actively trying to organize lithographers, just as other innovations have brought the process into plants where other unions have plant-wide bargaining units. In order to protect the product market, the ALA is forced to spend much time and money attempting to reorganize the organized. The roles have been reversed, and the ALA is now using the same tactic which, when used against it by the IPP&AU during the 1920's and 1930's, was so resented by the ALA.

The ALA: Policies and Problems

The ALA was fortunate in having exceptionally able men to determine strategy for the union from 1930 into the 1940's. Andrew Kennedy, president from 1930 to 1939, shifted the union away from a view of "Mr. Employer" as an enemy to be conquered, and toward the view that both union and employer had roles in eliminating wage competition, and in cooperating against unions and employer associations in the printing industry.

With President Kennedy's death, the responsibility for guiding the union fell to General Counsel Benjamin Robinson. He continued Kennedy's role as architect of industry cooperation, and in addition was successful in interposing the NLRB as a shield against raiding by the printing unions. The decision in the *Foote and Davies* case specifically rejected the historic AFL decree granting jurisdiction over offset to the IPP&AU and IPEU. The grounds on which the decision was reached were used by the ALA to establish in addition the right to follow the process into established bargaining units. This extension of NLRB protection was secured shortly before the AFL in effect replaced the principle of paper jurisdiction with the principle of jurisdiction in fact. The ALA continues out of step with the AFL, whose no-raiding agreement supports the ALA position prior to 1946 but not the position of the ALA today, since the "traditional lithographic unit" is now found in many firms which are organized by other unions.

Since 1946, preserving the separateness of the lithographic industry has not been a subject for union-management cooperation. This is so partly because many lithographers now are in combination shops, but it is primarily true because employers no longer feel that there is much to be gained by such separateness. In 1946 the international was forced by its largest local to adopt an aggressive bargaining strategy, which was so outstandingly successful in securing immediate gains for the membership that it became the fixed policy of the union. The employers have been unable to organize resistance nationally to union pressure, but since that time they have also been unwilling to seek areas of cooperation. The ALA, long familiar with opposition from the rest of the labor movement, has reentered a period in which employers also oppose it.

Yet it appears certain that the union cannot prosper unless it either has protection from the attacks of rival unions or support from employers. At present the ALA is at the crossroads and has not decided which road to choose.

On the one hand it seeks to reestablish cooperation with employers, on the other it continues to look for outside trade union support. Its leadership is impressed with the longrun benefits of technological change and would like to see present employers take advantage of the innovations that are currently available and cooperate with the union in speeding their introduction. Since it is the only organization with a significant degree of power in the existing industrial relationship, some of the leaders of the ALA have come to recognize the value of an effective national employers' association that is dedicated to the advancement of the lithographic industry. The interests of such an association and of the union would coincide on many points. In the 1920's when the NAEL was still a power in the industry, the secretary of that association complained that "some of the worst offenders in the matter of training apprentices are the biggest kickers when they cannot get a skilled man." [1] In 1958 such employers had become the union's problem:

> The thing that makes me maddest, a guy calls me — "Gee, what a bum you sent me." So I ask him, "How many apprentices have you trained?"
> "Well, not many, but . . ."
> "Well, where the hell do you get off asking for the cream?" [2]

One of the few ways in which the union can counter the spread of the process into firms beyond its immediate control is to ensure that present employers are not limited in their expansion by a shortage of skilled labor. But against this, as in its support of technological change, the union leadership is forced to weigh the historic goals of restricting the supply of labor and preserving the *status quo* in the industry. Although the ALA would like to see a national employers' association able to defend the industry view forcefully and to cooperate in areas of common interest, it has by no means regained the degree of commitment to employer cooperation that guided Andrew Kennedy's thinking.

Paralleling its partial attempts to reestablish cooperation with employers, the ALA seeks to protect its flank by allying itself with other unions. Currently attention is centered on a working agreement with the photo-engravers and the ITU. Such agreements require flexibility and respect for other viewpoints, two qualities which seem in short supply in the unions today.

The unions may of course continue to go it alone, though costs here are high and continue rising. Present per capita cost of international salaries and organizing expense in the ALA is double the level of 1948, and this increase does not take into consideration local administrative costs. Dues in the larger locals run from $8 to $10 per month, which represents a substantial slice of take-home pay. The ALA has a measure of "free" protection as long as the NLRB continues to follow its present policy of carving out lithographic units, but such a policy is subject to change, and no one is more aware of this than the ALA leadership. In late 1958 the ALA was debating whether to revise the whole organizing approach, to give up the "all lithographers and only lithographers" policy, and to expand on either a craft or industrial basis. This is not the first time the question has arisen, but not since 1946, when the ALA became the only graphic arts union in the CIO, has a basic policy shift been given such serious consideration.

The Future Course of Bargaining

Innovations, which have given lithographers so lucrative a present and so uncertain a future, are going to continue to be introduced. Phototype-setting is inherently more economical than hot-type composition, both because the raw material used is less expensive and because the necessary operations (other than keyboarding) are more easily mechanized. Automatic color separation is in its infancy; it is likely that the initial breakthrough in separating colors electronically will soon be followed by the elimination of separate color correction and half-toning steps.

Perhaps of all innovations, the plastic relief plate is the most important since it will reduce letterpress make-ready to operations which are identical with those in the lithographic process. Both processes will use type set photographically with maximum efficiency; both will use the same type of color separations and will require the same stripping operations and closely similar platemaking operations. If, as is likely, the plastic plate runs best if its printing surface is protected from the abrading effect of paper, then it will be run on a dry offset press, a lithographic press without the dampening rollers.

It is inevitable that, with developments of this type in the graphic arts, the distinction between lithography and letterpress is going to be reduced to the running qualities of different types of plates on essentially the same press. This will hold true for substantially all printing markets, and probably also for newspaper printing. It seems reasonable to conclude that the city-wide bargaining pattern common in the lithographic industry will become increasingly unstable because it is based on a process distinction. An increasing tendency for bargaining units to follow product market divisions will develop. This will accentuate the identity of the ALA's interests and problems with those of the letterpress unions, which already have such a bargaining pattern. It is unlikely that it would permit a return to national bargaining for that segment of the industry oriented to the national market, since this market is served by so many different types of firms. These firms will band together into a strong organization only in the unlikely event that the ALA has some remarkable success in organizing lithographers away from other unions in captive plants, and can apply such pressure that the employer's self-interest is best served by setting up a strong national association.

It is also evident that the mobility of the offset process will further reduce the independence of wage-making in lithography from wage-making in the rest of the printing industry. Put more generally, there will be a tendency for the basic wage contour in the lithographic industry to lose its separate identity, and to become a feature of the basic wage contour in the graphic arts industry. This will be a big step toward the merging of the lithographic industrial relations system into a new graphic arts system. It has been shown that such a trend is evident at present; there is much to indicate that it will continue.

The Future Structure of Printing Unions

Such developments point to the further possibility that the structure of unionism in the printing industry will change. The historic reason for a separation between lithographers and other printers will be eliminated, and with it the reason for a separate lithographic union. Without a market distinction, without a process distinction, and without a separate group of skills, the ALA will find it exceedingly difficult to continue as an independent policy-making unit, though it could perhaps maintain nominal independence by keeping in line with other unions.

Other printing unions will also be affected. The electrotyper and stereotyper crafts may be eliminated and the photo-engravers substantially reduced in importance. Five key skill classifications may remain — the typesetters, cameramen, strippers, platemakers, and pressmen. The question arises then of the probable structure of unionism in the printing industry if technological change does in fact lead to a blending of the processes. Historically, union organization in the printing industry has been determined by three variables — the nature of the market, the structure of the industry, and the skills required by the process. Technology has been the central feature in determining each of them. For example, the linotype was too productive, too expensive, and its productivity too dependent on steady use for it to be a good investment for most commercial printers. The market was rather lax in its demand for promptness. An order delivered twenty-four hours late was unfortunate, but hardly a disaster of the magnitude of a newspaper delivered twenty-four hours late. Moreover, it was quite feasible to separate typesetting from the rest of the printing process, and the trade typesetting shop was an inevitable result. For essentially similar reasons photo-engraving also became predominantly a trade shop operation. To a lesser degree the making of electrotypes added a third variety of trade shops to the printing industry, and bindery work a forth.

In each case industrial structure changed. This change did not, however, reduce the demand for skilled craftsmen. With the possible exception of bindery work, there continued to be a key skill in each operation, which required time to master and could not be replaced readily by another skill or by a machine.

In lithography the cost and minimum efficient output made possible by innovations did not outrun the absorptive capacity of the firm of average size. Where they did, the greater integration of operations which the chemical nature of the lithographic process required made trade shop operations less feasible. There is little question that the splintering of the printing operations in letterpress contributed to the splintering of one process-organized union into five craft-organized unions. Similarly, there is little doubt that the face-to-face contact maintained by the integrated process in lithography helped bring five craft-organized unions into a single process-organized one.

Once established, the structure of unionism in an industry takes on non-economic justification, and it is not to be expected that it will shift with every shift in the nature of the market, the structure of the industry, or the skills required by the process. But the shift which is taking place in the printing industry today is fundamental, and it is reasonable to believe that there will be some realignment among the worker organizations.

In one respect the key to the puzzle lies with the ALA. It is the only graphic arts union which is well organized in three of the four basic classifications, and if it joined with the ITU, one union would have a solid membership base in all classifications. A union with the IPP&AU, on the other hand, would give one union control over all organized pressmen in the printing industry. A union with the IPEU would make both unions stronger in ways in which they are already strong, but would not add a new element to the power of either. A union with the bindery workers would strengthen both unions in lithographic plants somewhat, but would probably embarrass the latter union in its relations with the letterpress unions in other plants. Viewed solely as a question of bargaining power, preference should rest with the IPP&AU, the ITU, and the IPEU in that order; as a practical matter, probabilities run in precisely the opposite direction.

It is entirely possible that interunion conflict will have to become far more acute than at present before there will be sufficient member interest to permit amalgamation. There is a practical problem of finding men with the prestige, ability, and time necessary to work out the serious issues which still divide the unions. Men with the necessary prestige are national officers; if they have the ability necessary, they are already heavily burdened with the more immediate problems of organizing, negotiating, and carrying out the details of administration which a democratic organization requires of them. Points of irritation constantly arising between the unions will require immediate and tactful handling if an atmosphere is to develop that would permit a favorable membership vote on amalgamation. Yet these are frictional and not fundamental difficulties. They are no greater than those which faced the lithographic craft unions in 1915. Leaders may arise who will speed amalgamation, as Robert Bruck was able to do in 1915, or delay it, much as John L. Lewis and others delayed the reunification of the AFL and CIO. Conceivably the ALA could go the way of the poster artists, unwilling to compromise until the changing economic setting eliminates all possibilities except becoming a part of another union.

The ALA's choice of leaders will affect its own fortunes, but will have little effect on the structure of union organization in the graphic arts. Like the organization of the industry and the pattern and results of bargaining, the structure of unionism will in the longrun adapt itself to the context of the industrial relationship. One union may decline or grow strong without affecting the structural requirements for effective unionism in the industry.

Some Generalizations Concerning Change in an Industrial Relations System

The direct and most evident conclusions of this study apply to the subject matter of the investigation. Still, certain more general conclusions concerning the nature of working rules and the rule-making relationship have been suggested. It is clear, for example, that the reasons for rules in most cases can be traced back to one or more parts of the context of the industrial relations system. Rules concerning apprenticeship ratios or the manning of presses are a direct response to the pace of technological change interacting with changes in the demand for the product. Apprenticeship itself is a response to skill requirements enforced by production technology.

The sources of other rules are not so obvious. For example, it was once commonly accepted in the lithographic industry that the setting of wages and working conditions was a national question. This rule, for such in fact it must be considered, was based most directly on a uniform technology and a homogeneous group of workmen within the industry who rapidly formed a strong national union. But rule-making continued at a national level when the union was eliminated from the rule-making process, partly to ward off the union, but also because key employers shared a common product market. During the 1930's under the NRA the practice received formal encouragement by government, but by 1937 the rule was decisively discarded, in spite of the best efforts of both the union and the employers' association. One must conclude that the key to this rule was the existence of a national market for lithography. With the expansion in the number of different product markets, national rule-making was no longer feasible.

The important fact is that by a study of the context of an industrial relations system the explanation for rules of the system can be identified. It may be discovered that the initial reason for introducing a practice is not at all the same as the reason for its current continuation, but this itself is important information since only with such facts can one measure the possibility of change, or suggest viable alternatives.

The responsiveness of working rules to pressures for change. Some rules in an industrial relations system are highly responsive to a change in the context in which the system operates. Certain changes in the product market, for example, which strengthen or weaken the unity of employers, or materially alter the losses incurred as a result of a strike, will have a prompt result in wage decisions. Again, the wide-spread introduction of a new method which decreases the demand for a craft skill will be promptly followed by a change in the rules concerning the making of apprentices.

Other rules are less responsive. The same change in the product market which led to a prompt change in wages may also be a pressure for different bargaining arrangements, perhaps shifting from a national to a city-wide basis or from a city-wide to a product-market basis. But existing bargaining arrangements will change only slowly. Again, changes in the production process or in the product market may make different organizational arrangements within the union desirable, as for example a shift from craft to regional representation, but such changes will also be made only slowly. Other examples could be given; each would illustrate one or both of the following truths: (1) the rule involved was a response to other parts of the context in addition to the part of the context being changed, and (2) the rule itself was a basis upon which other rules had been formed. Bargaining arrangements, for example, imply negotiating bodies for both sides, and procedures for selecting and controlling them. In general, rules which are necessary to make the system operate, which are in effect a part of the framework of the system, will be less responsive to change than those which apply directly to the workplace.

Some rules are unresponsive to any change in the context of the system. Such, for example, are the rules which are necessary for the security of any actor in the system. No change will be made in the rules against scabbing or in the rules enforcing governmental edicts. It is not necessary that the rule in fact protect the security of the organization, only that it be believed to have that effect. Does a person have to serve his apprenticeship in the trade to be a good union officer? The fact of the matter is less relevant than the belief by the great majority of the membership that this is so. That it is not true in all trade union movements, or that the union of pressmen and transferrers (the LIP&BA) had a feeder for their president within fifteen years after amalgamation, is not germane. The rule, the genesis of which is to be found in the status of the union in an earlier age, now has a rationale of its own in the minds of members.

Other pressures will also make rules unresponsive to changes in the context. Where there is uncertainty as to the results of a change in the rule, or where emotions are heavily involved (the two are often related), the rule will be exceedingly difficult to change. It will make little difference that from a rational point of view the rule may be minor. Examples of such rules are not uncommon. The rule against any pressman running two presses, and the blanket requirement of a four-year apprenticeship for all crafts are two examples in this study.

It is important for anyone interested in varying the rules of the workplace to be acutely sensitive to the differing degrees of flexibility in rules. Whether the desire is to increase union-management cooperation, to improve the workingman's lot, or to provide more effective national policy, consideration must be given to the reasons for rules, both current and historic, in order to measure the resistance which will be encountered when attempting the change.

Some considerations when modifying the rules of an industrial relations system. The above propositions concerning the responsiveness of rules to changes in the context of the system have certain corollaries of more direct application.

(1) A rule which is based on one part of the context will not respond readily to a change in another part of the context. For example, the offer of a million dollars by the ALA to cooperate with the employers in advancing technological change was not likely to cause any change in the firm resolve of the national employer associations not to have direct dealings with the union. Direct contact had been broken off when it became evident that at the national level the power of the union far outweighed that of the employers. The million dollar offer, if accepted, would have increased the power of the union rather than have provided a power balance.

To take an example on the employer side, the exhortations common during the late 1940's that employers ought to stiffen their backbone in dealing with the union were meaningless, since the problem was not in morale or the state of their convictions, but in a radically changed market structure. Such exhortation had a legitimate role in the first thirty years of the century, and unquestionably the *esprit de corps* of the NAEL membership contributed to the very real strength of that association during these early years.

In the case of government, the prohibition of closed-shop clauses under the Taft-Hartley Act almost seemed to assume that power sprang from the contract, and not the contract from power. If the rationale of union hiring-hall arrangements is to be found in the nature of the labor market and not in the working of the labor agreement, then it is idle to assume that any change which does not operate on the labor market will cause a substantive change in the hiring hall.

(2) Some rules in an industrial relations system are highly interdependent. A change in one rule will force (or permit) a change in another. Examples immediately suggest themselves. The government decision to give the National Labor Relations Board the right to determine bargaining units caused the ALA to centralize all decisions relating to jurisdictional questions in the office of their general counsel. The introduction of the thirty-five-hour week led to more stringent rules concerning moonlighting in the trade, and overtime authorization.

Particularly when one is interested in introducing change from the outside, as for example by changes in the law, it is useful to be aware of this interdependence. An industrial relations system is a system, and not several sets of autonomous rules for wages, for working conditions, for bargaining arrangements, for organizing, and so forth.

(3) Some rules, believed to be inflexible, are changed with remarkable ease once the nature of their strength is clearly understood. One of the clearest examples of this is found in the operation of closed, or at least tightly insulated, local lithographic labor markets. In such markets, where the union acts as the disburser of labor skills, certain supposedly sacred rules are largely ignored by the workforce. Individual bargaining, so generally conceded to be in the worst interests of the trade unionist, is not only approved but encouraged and even arranged through the trade union. It is through such bargaining that premiums above scale are made possible. These premiums may and quite often do upset another time-tested belief, that nothing creates so much unrest in a workforce as the reversal of com-

monly accepted wage relationships. But as we have seen, it is possible through the payment of premiums to do precisely that, and with no apparent ill effects. Again, the classification of seniority provisions as a "bosses' rule" is an unusual classification by trade unionists. Yet on inspection it is clear that each of these three rules — the prohibition of individual bargaining, the maintainence of wage differentials, and seniority protection — are intended to protect the worker from the ill effects of uncontrollable pressures in the labor market. When the labor market itself is brought under control, these three rules, each of which interferes with an optimum distribution of labor, go by the board.

Relatively little thought has been given to appropriate public policy in such cases. Yet it is important, for such labor markets are in existence in many industries in many localities, operating at unknown but probably low levels of efficiency. It must be remembered that they operate at the outskirts of the law, and are operated by individuals without formal training yet deeply experienced in the traditions of trade unionism. Much might be done to improve the efficiency of such markets.

But is the existence of such closed markets desirable? Little beyond Clark Kerr's "The Balkanization of Labor Markets"[3] has been written of the question, yet few questions are of greater significance to the efficient operation of labor markets. One could argue that the demonstrated failure of the state employment services to eliminate labor market imperfections justifies encouragement of this piecemeal approach, eliminating laws against it and instead providing laws or services which ensure adequate entry to the market and efficient management of the labor marketing service.[4]

A study of a single system such as this throws some light on the problem, which of course requires a broader analysis for its solution. There is evident value in an approach that explicitly recognizes the existence of an *industrial relations system,* quite separate from the economic system that gives to the firm the role of rule-maker. For a continuing theme of this study has been that the context of the lithographic industrial relations system provides a setting in which several of the actors interact in producing the rules of the workplace. For some rules the context of the system will give to one actor a dominant role, for other rules another of the actors will dominate this decision process. For some rules the government may be able to influence results by operating directly on the other actors, but to change other rules it may be required to operate on the context itself.

Implementing public policy in industrial relations: a final illustration. As unions have increased in power, there has been a parallel increase in the concern that this power will be used to the detriment of the nation and the diminution of individual member freedom. Some have urged that our public policy has made insufficient use of the free play of competition to restrain the ill effect of this power. Perhaps only the obscurity of the lithographers union has prevented such people from using that union as an example of the many benefits obtainable from competing unionism. They are democratic, they support technological change, they do not suffer from low wages or high hours, or indeed from unemployment, yet their jurisdiction has been under constant challenge by other unions for well over forty years.

The extreme superficiality of such a conclusion is apparent when one considers the impact which innovations have had on broadening the market, which the multi-craft structure of the union has had in permitting the adoption of innovations without displacement of members, which the NLRB has had in providing a protective shield to bargaining units, which the market enforced reluctance of employers to take a strike has had on wage gains, which the Socialist heritage has had on the distribution of authority within the union, and on and on. Competition from other unions has also had an effect on keeping the ALA responsive to its member needs, but one only need recall that the ALA membership, in comparison with the letterpress trades, was worse off before the mid-1940's and better off thereafter to recognize that competition between the unions did not even ensure wage equality, and could scarcely be the prime cause of leadership responsiveness.

Surely the matter of leadership responsiveness to member sentiments is a legitimate concern of public policy, but it does not follow that such responsiveness will be kindled by a new election procedure or the encouragement of dual unions. Forty years ago R. F. Hoxie said the reason for studying trade unions was to learn ". . . what ought to and can be done to control them." [1] His choice of verbs was excellent. Policy makers should not carry the concept of a New England town meeting in their minds as the model for the internal life of a trade union, or two merchants competing for customers as a model for the way unions should act in the labor market. Apart from their inconsistency with each other, neither can be achieved.

What ought to be achieved may well be impossible to legislate directly. A more useful approach would be to consider possible changes in the context of the system itself. To take but one example, a recognition of the permanent status of unions might well be reflected in a greater concern for the training of trade union leaders, though not necessarily so great as is presently reflected in our many tax-supported business schools. Another, less indirect effort would be to provide for coordination of governmental and union employment services, or in other ways to encourage labor mobility. For high mobility encourages leadership responsiveness, as well as providing other economic benefits.

Problems of leadership responsiveness in the trade union movement, no less than other aspects of the "labor problem," can best be understood if they are viewed in the context of the system in which they arise. With this understanding, there is greater hope that policy makers can determine what ought to, and can be done about them.

Appendices

Appendix **A** ✆

Lithographic Industry Statistics

Definition

The Bureau of the Census accepts the standard definition of the lithographic industry:

> This industry comprises establishments primarily engaged in printing by the lithographic process. The greater part of the work in this industry is performed on a job or custom basis; but in some cases lithographed calendars, maps, posters, decalcomanias, and so forth, are made for sale. Offset printing, photo-offset printing, and photolithographing are also included in this industry. Establishments primarily engaged in lithographing books and pamphlets, without publishing, are classified in Industry 2732, and greeting cards in Industry 2771.

This definition has changed over time, notably in 1929, 1939 and 1958. The effect on comparability of data is discussed in notes [e], [f], and [g] of Table 20.

Table 19. Statistics of the printing and publishing industry.
(*S.I.C. 27—Printing and Publishing*)

Year	Establishments (thousands)	All employees (thousands)	Production workers (thousands)	Average annual wages	Value added (millions)
1899	24.36	244	202	510	300
1904	28.37	299	228	580	424
1909	32.14	374	270	640	519
1914	34.24	406	286	720	627
1919	33.26	448	304	1150	1090
1921	22.56	428	284	1530	1300
1923	22.90	481	310	1590	1530
1925	23.65	497	317	1710	1760
1927	25.3	524	331	1780	1940
1929	27.4	566	358	1780	2230
1931	24.7	—	315	1700	1770
1933	19.2	399	262	1350	1250
1935	22.5	472	303	1470	1550
1937	22.7	555	351	1500	1780
1939	24.9	552	324	1520	1770
1947	29.0	715	438	3000	4250
1954	32.5	804	500	4230	6260
1958	not available	864	527	4900	7795

Source: 1899–1954: *1954 Census of Manufactures*, vol. II, p. 27B-1; 1958: *1958 Census of Manufactures*, MC(P)1 (December 1959), p. 7.

Table 20. Statistics of the lithographic firms in the printing industry.
(S.I.C. 2761—Lithographing)

Year	Number of establishments[a]	All employees (thousands)[b]	Production workers (thousands)[c]	Average annual wages[d] (in dollars)	Sales volume (millions of dollars)	Value added (millions)
1849	11	—	.16	315	.14	.09
1859	53	—	.79	430	.85	.62
1869	91	—	1.4	600	2.5	1.8
1879	167	—	4.3	540	6.9	4.2
1889	219	—	9.7	600	17.9	11.7
1899	263	—	13.0	530	22.2	14.4
1904	248	14.2	12.6	650	25.5	16.9
1909	318	17.1	15.1	675	34.1	22.2
1914[e]	336 285	19.0	16.0	780	41.0	26.1
1919[e]	331 290	20.9	17.1	1165	80.2	49.1
1921[e]	325 313	19.0	15.2	1510	85.1	52.1
1923[e]	350	22.0	17.8	1550	99.2	64.7
1925[e]	352	23.0	18.6	1600	106.8	70.9
1927[e]	337	22.2	18.1	1690	106.9	69.6
1929[e,f]	372	24.8	20.1	1690	127.8	86.2
1931[f]	352	19.4	15.7	1590	84.4	58.8
1933[f]	328	16.5	13.8	1250	64.6	40.0
1935[f]	362	20.3	16.5	1370	85.9	54.5
1937	516	29.8	22.5	1500	129.2	78.5
1939	789	34.7	26.9	1460	159.5	99.9
1947	1415	52.4	41.4	3080	488.6	314.1
1954	2924	77.1	60.2	4330	962.6	578.9
1958 (old basis)	4259	97.5	74.1	5150	1473.5	872.4
1958 (new[g] basis)	3733	89.3	67.9	5250	1337.7	794.9

Sources: 1849–1921: *1921 Biennial Census of Manufactures*, p. 657; 1921–1929: *Census of Manufactures, 1929*, vol. II, p. 612; 1931–1958: *Census of Manufactures, 1958*, vol. II, p. 27B-4.

[a] Number of establishments
　　1849–1921, all firms reporting
　　1914–1939, all firms with $5000 or more annual sales
　　1947–1958, all firms employing one or more persons at any time during the census year.

[b] Excludes working proprietors of unincorporated firms.

[c] Prior to 1947 the classification was called "wage earners," but no material change has been made in the definition. The classification excludes supervisory, office clerical, sales, and technical (professional) employees.

[d] This column is simply the total wage bill divided by total wage earners. The calculation was made prior to adjustments described in notes (e) and (f) below.

[e] 1914–1929: All figures except Average Annual Wage adjusted to include an estimate for lithographic firms classified in "Labels, Tags and Seals" for those years. Estimate, based on explanation in *1931 Biennial Census*, p. 526, is 29 per cent of "Labels, Tags and Seals" figures. In all cases adjustment was less than 10 per cent.

[f] 1929–1935: All figures except Average Annual Wage adjusted to exclude an estimate for lithographic firms later (1939) classified in "Greeting Cards." The 1937 data were recalculated by the Census Bureau. Estimate based on proportion of greeting card sales volume to total sales volume for each year, and other figures reduced by the same percentage. In all cases adjustments were less than 7 per cent.

[g] In 1958 a new industry, "manifold business forms," was carved out of the letterpress and lithographing census. In addition, establishments producing playing cards and certain lithographed tags were transferred from the paper products to the lithographing industry, while others producing decalcomanias by letterpress, gravure or screen process were transferred from the lithographic to the letterpress industry. The Bureau of the Census provides basic data on the old classification system, but more refined calculations (establishments by size class, specialization and coverage ratios, and so forth) are based on the new classification.

Table 21. Statistics of the letterpress industry.
(*S.I.C. 2751—Commercial Printing*)

Year	Number of establishments[a]	All employees (thousands)[b]	Production workers (thousands)[c]	Average annual wages[d] (in dollars)	Sales volume (millions of dollars)	Value added (millions)
1939	10,295	—	112.4	1,410	607.8	381.9
1947	11,920	191.7	154.5	2,900	1,513.1	970.3
1954	12,073	200.2	157.6	4,130	2,202.1	1,357.8
1958 (old basis)	13,150	199.3	156.3	4,810	2,722.6	1,591.0
1958 (new[g] basis)	13,002	189.1	149.1	4,800	2,533.2	1,482.6

For notes and sources see Table 20.

Table 22. Distribution of firms by size, *1919–1954*.
(*S.I.C. 2761 — lithographing*)

Year	Classification	Size of firm by number of wage earners				
		1–20	21–100	101–500	Over 500	Total
1919	Establishments	139	113	34	4	331
	Wage earners	1,210	5,240	6,400	2,780	15,618
1929	Establishments	171	154	45	1	376
	Wage earners	1,600	7,000	10,400		18,979
1939	Establishments	469	201	62	3	749
	Wage earners	3,660	8,900	11,400	2,000	26,000

Year	Classification	Size of firm by number of employees				
		1–19	20–99	100–499	Over 500	Total
1947	Establishments	923	363	119	8	1,413
	All employees	6,000	15,100	22,900	8,200	52,400
1954	Establishments	2,156	598	158	12	2,924
	All employees	13,800	26,000	30,000	7,900	77,717
1958[a]	Establishments	2,720	1,013	166	12	3,733
	All employees	16,957	33,656	30,504	8,188	89,305

Sources: 1919: *Fourteenth Census of Manufactures, 1920,* vol. VIII, p. 87; 1929: *Census of Manufactures, 1929,* vol. I, p. 96n; 1939: *Census of Manufactures, 1939,* vol. I, pp. 141–143; 1947: *Census of Manufactures, 1947,* vol. II, p. 369; 1954: *Census of Manufactures, 1954,* vol. II, p. 27B-12; 1958: *Census of Manufactures, 1958,* vol. II, p. 27B-11.

[a] Data for new industry classification; not strictly comparable to previous years. See note g following Table 20 of this Appendix.

Estimates of the Lithographic Workforce and its Representation by the Amalgamated Lithographers of America

I have been given access to information by both union and industry sources, with the entirely justified request that it not be used to the disadvantage of the person or organizations providing it. Some researchers prefer to be free from such restraints, but I have chosen to accept the limitations to get the material. Some of this information has proved of value in the making of the estimates contained in this Appendix. Naturally, it has been necessary to avoid citing sources, or even using the data in a concrete way. The estimates are not precise, but they are not entirely the exercise in imagination they may appear to be.

I. *How many workers are there in the census-defined lithographic industry?*

Three government agencies collect statistics on employment by industry and by occupation — Bureau of Labor Statistics, *Monthly Labor Review,* appendices; Industry

Table 23. Employment in the lithographic industry. (thousands)

Year	All employees			Production workers		
	BLS	Census of manufactures	Estimate[a]	BLS	Census of manufactures	Estimate[a]
1947	49.5	52.4[b]	52.4	39.3	41.4[b]	41.4
1948	50.8	n.d.	55.5	39.6	n.d.	44
1949	50.4	n.d.	59	39.1	n.d.	46
1950	51.8	52.6	62	40.4	38.5	49
1951	53.9	55.1	65.5	42.0	40.4	51.5
1952	54.6	57.0	69.5	42.2	42.3	54
1953	57.7	57.3	73	44.6	45.0	57.5
1954	60.5	77.7[b]	77.7	46.4	60.2[b]	60.2
1955	62.0	79.8	82	46.9	61.7	64
1956	64.3	81.4	86	48.5	62.5	68
1957	66.7	83	91	50.7	64	71.5
1958	65.7	97.5[b]	97	49.7	74.1[b]	75
1959	66.3			50.1		

[a] Based on constant rate of growth, using only 1947 and 1954 figures.
[b] Total census. Other figures based on sample surveys.

Table 24. Estimated size of the lithographic workforce, 1958.

Category	Number
Lithographic firms in printing industry	58,700[a]
Metal decorating	3,500[b]
Captive paper plants	7,000[c]
Private plants	3,000[d]
Combination plants	15,500[e]
Publishing houses	17,600[f]
	105,300

[a] Lithographic firms in the printing industry: Only 79 per cent of the lithographic industry's production was lithography. The workforce figure is reduced in proportion.

[b] Metal decorating: firms are not reported separately in the Census. The estimate here is based on interviews with ALA and National Metal Decorators Association officials. It includes both captive and independent shops.

[c] The paperboard industry (S.I.C. 265, 266, 267, and 269) employed 28,164 workers engaged in printing of all kinds in 1958. See *1958 Census of Manufactures*, "Special Report on Commercial Printing," MC(P)-27B-Supplement (October 14, 1960), p. 2. At a conservative estimate, 25 per cent of all printing in this industry was by the lithographic process in 1958, and this is the basis for the figure in the table. But see also the employer brief in the *Sutherland Paper Company* case, NLRB 7-RC-3793, 1958, for data which would give a much lower estimate.

[d] Private-plant employment is largely unknown. If all Multilith operations were included, the figure might reach 100,000 workers. But such workers should not be included, and this limitation reduces the figure to perhaps 5 per cent of the lithographic industry workforce. Nothing is claimed for accuracy, but it is believed that inclusion of the estimate on balance improves the accuracy of the total estimate.

[e] Combination plants: lithographed products are produced in many industries, primarily the commercial printing industry. The device used here is to divide lithographic sales volume produced *outside* the lithographic industry ($281,000,000) by the average sales volume per worker *within* the lithographic industry ($19,900). This figure (15,500) is taken as the number of production workers in combination plants.

[f] Publishing houses: this estimate is based on individual calculations for the industries below.

For census purposes, book printing is divided between firms which print and publish, and firms which print for publication by others. In the latter product group, 40.1 per cent of the $417 million sales volume was lithographically produced. If this percentage is applied to the production workers in both industries (thus avoiding the complications of compensating for receipts attributable to publishing), the combined workforce (11,000 in printing and publishing and 23,000 in printing only) contains 14,000 lithographic production workers. However, the workforce in these industries contains a higher proportion of unskilled workers than is true for lithographic firms generally, a fact which is reflected in lower average annual wages ($5140 in lithography, $4040 in publishing and printing). Adjusting the estimate to reflect this fact, the lithographic workforce would be 10,500 workers.

The greeting card industry (S.I.C. 2771) can be handled in much the same way. Of greeting cards printed for publication by others, 48 per cent are produced lithographically. The workforce of 14.6 thousand workers is then presumed to contain 7000 working on lithography.

The 14,600 workers, however, include many unskilled workers, who pack, sort, and otherwise prepare the cards for shipment. This workforce distribution is evident in the fact that the average annual wage is $3420, by far the lowest in printing and publishing, and 20 per cent under the national average for manufacturing. A trade estimate has placed the number of printing workers at 7500, which is accepted here, giving a lithographic workforce estimate of 3700.

The periodicals industry (S.I.C. 2721) has 11.5 thousand workers, but no figures to indicate the proportion of work produced lithographically. Many magazines are produced outside the industry, however, and in this case a breakdown is available. The value of such production was $729.8 million in 1958, of which $87.2 million was produced lithographically (see source cited in Appendix C). Applying this percentage (12%) to the workforce gives a figure of 1400 workers.

Although lithography is expanding rapidly in the newspaper industry it is still a comparatively insignificant contender with letterpress. This and other miscellaneous publishing account for some 2000 additional lithographic workers.

The addition of those estimates for lithographic workers in publishing of all kinds gives an estimate of 17,500 lithographic production workers.

Division, Bureau of the Census, *Annual Survey of Manufactures;* Population Division, Bureau of the Census, Preprint Series, *Occupation by Industry.* The Amalgamated Lithograpers of America also collects statistics on its membership, and it is from these four separate sources that basic employment estimates must be drawn.

Industrial employment figures are reported monthly by the Bureau of Labor Statistics, and annually by the industry division of the Bureau of the Census. Both sets of figures are based on sampling techniques. Industrial employment figures based on a total census are available for the years 1939, 1947, 1954, conducted by the latter agency.

Both these agencies gather their employment data on the industry definitions provided in the *Standard Industrial Classification Manual,* for "all full-time employees . . . who worked or received pay for any part of the pay period ending nearest the fifteenth of the month," yet the results vary widely. Figures from two sources are given in Table 23. The Census figures are considered more reliable because they depend on a complete coverage, and for intercensal years the assumption of a constant rate of growth is used rather than either the intercensal annual surveys or the BLS figures. If BLS figures were uniformly below or above the Census figures, it might be possible to make some use of them, but as it stands they seem largely useless.

The annual surveys of the Bureau of the Census are obviously unreliable, apparently because the data-gathering procedure used has a built-in tendency to ignore the crossover of firms from letterpress to lithography.

II. *What workers from outside the industry should be included?*

For the purposes of this study, the "workforce" is not limited to lithographic firms in the printing industry, nor are all workers in these firms included.

III. *What proportion of these 105,000 production workers are in the apprenticeable lithographic trades?*

The ALA does not bargain for all workers in a lithographic plant. Among those usually excluded are janitors, maintenance crew, paper cutters, floor help, trimmers, bindery workers, shipping and warehouse workers. In large integrated houses, this group of workers will outnumber lithographers two to one, although in trade shops the proportions will be reversed. Small firms in the industry commonly have trade binderies do their finishing work, and thus the proportion of lithographers to non-lithographers tends to rise in smaller shops. No Census breakdown is available to identify the number of nonlithographers, nor do industry sources provide useful information. In the absence of these, estimates are made directly of the number of workers in the different crafts. The attempt to provide a crosscheck of the size of the skilled lithographic workforce by subtracting an estimate of the nonlithographic workforce from the total 105,000 production workers has been discarded.

Occupational Distribution

The ALA conducted a survey in 1956 which revealed distribution of journeymen and apprentice members shown in Table 25.

The ALA is less well organized in the small shops than in the large ones, and its coverage of small press equipment is considerably less complete than its coverage of multicolor equipment. This is evident from Table 26.

Table 26 does not help determine the number of production workers operating offset duplicators. James Wilkinson in the December 1958 *Modern Lithography* placed

Table 25.   Occupational distribution of craftsmen in the ALA in 1956.

Classification		Percentage distribution
Preparatory department	—	28
Camera	7	—
Copy correction and engraving	10	—
Stripping	11	—
Platemaking department	—	12
Press department	—	29
Multicolor pressmen	11	—
One-color pressmen	14	—
Offset duplicator operators	1	—
Metal decorating pressmen	3	—
Feeders	—	17
All others	—	14

Source: ALA, *1956 National Wage Survey*. The union had a 76 per cent response from journey-men members, 69 per cent from the membership at large.

the number of such presses in operation at 90,000. Since there were only 105,000 production workers connected with lithography in 1958, it is apparent that very little of this equipment is in commercial printing firms of any sort. The same magazine carried an article in September 1958 by an officer of American Type Founders, a large manufacturer of small presses, who indicated that his firm and others found their initial market for this type of equipment among "the bedroom printers and the captive shops." It appears that such operations have remained the major buyers.

The foregoing paragraphs have indicated the sources from which occupational data

Table 26.   Distribution of offset presses and ALA pressmen.

Press	Press sales 1947 to 1956		Press equipment Survey, 1952		ALA pressmen (percentage)	
	No.	(per cent)	No.	(per cent)	1952	1956
1-color press under 30″	1829[a]	42[a]	1030[a]	45[a]	34[b]	57[b]
1-color press 30″ and over	1571	36	750	33	24.5	
2-color press	825	19	403	17.5	31.5	31
3, 4, 5-color press	185	4	105	4.5	6[c]	8[c]
Totals	4400	100	2288	100	96	96

Sources: Figures for press sales are for Miehle and Harris, the two largest manufacturers. *Press Equipment Survey* was done for the Lithographers National Association and the National Association of Photo-Lithographers by Dun and Bradstreet, Inc.; the figures are for January 1952. The figures for ALA Pressmen are from *National Wage Surveys* for the cited years. The 1952 figures are least reliable, as they have been drawn from preliminary figures for only 52 per cent of the membership. Both web offset presses and web offset pressmen are excluded from this table.

[a] Excludes offset duplicators.

[b] Includes offset duplicator operators.

[c] Includes first pressman only.

may be gathered, the absence of any information in some areas, and the difficulty of drawing the available material together to form a reliable estimate of the occupational distribution of the lithographic workforce. The list below makes use of the material presented above, but in view of its nature also draws on what seems reasonable to the writer after two years of research in the industry.

Category	No. of workers
Preparatory	15,000
Platemaking	5,500
Multicolor pressmen	4,000
One-color pressmen	11,000
All other pressroom crafts	10,000
All other lithographic journeymen and apprentices	3,500
Total journeymen and apprentices	49,000
Other lithographic production workers	56,000
Total production workforce	105,000

What Is the ALA's Degree of Organization?

The ALA would, when pressed, admit that there were some lithographers in other unions, and a somewhat larger number unorganized. It would still claim that it had at a minimum 75 per cent of all lithographers in the union. For several reasons this figure seems excessive. During the mid-1950's the ALA's organizing activities faltered seriously, and the number of unorganized lithographers grew at four to five times the rate of growth of the ALA. Moreover, the expansion of the process has been throughout so many industries that it is even difficult to find all lithographers, quite apart from organizing them. (The ALA leadership actively disputes the accuracy of the workforce estimates made here, believing that they seriously overstate the number of workers connected with the lithographic process.) If the figures in the previous list are accepted as accurate, however, the ALA's 29,100 journeymen and apprentices in 1958 represented slightly under 60 per cent of the total craftsmen. It is of course true that the 49,000 estimate includes small one- and two-man operations, and others which the ALA makes no attempt to organize. But the union has unquestionably lost ground since the late 1930's.

To make an estimate of the degree of organization by the ALA in major cities, the number of lithographic production workers in six standard metropolitan areas (as defined by the Census) was estimated for 1957 by projecting the 1954 figures. Seventy-five per cent of the 1954–1958 increase in lithographic production workers in the states in which the cities were located was the basis of the projection. These estimates appear in column 1, Table 27.

The number of journeymen and apprentices was calculated by assuming that in cities in which employment averaged from twenty-five to thirty persons per establishment, the workforce was divided evenly between apprenticeable and unskilled jobs. As establishment size increases, the proportion of skilled workmen is presumed to decrease, to a low of 35 per cent for cities with average employment per establishment of sixty or more. As average establishment size decreases, the proportion of skilled workmen is presumed to increase, to a high of 65 per cent for cities with average employment per establishment of fifteen or less. These estimates appear in column 2 of Table 27.

The number of ALA journeymen and apprentice members of ALA locals in each of these cities in 1957 was then calculated as a per cent of the column 2 estimate, and the result appears in column 3.

The accuracy of these estimates is low, for three important reasons. (1) The workforce figures are based on the Census-defined industry, which is not coterminous with the claimed jurisdiction of the union. (2) The determination of the number of skilled workmen is based on an arbitrary formula. (3) This monograph is being prepared two years after the *1958 Census* was taken, but data on standard metropolitan areas do not appear until three years after the raw data is collected. For this reason 1954 rather than 1958 data for standard metropolitan areas has had to be used.

Table 27. Estimates of ALA degree of organization in seven cities, 1958.

Standard metropolitan area	All production workers	Journeymen and apprentices	Proportion organized by the ALA (in per cent)
New York	13,000	6500	86
Chicago	9,000	4900	92
San Francisco	3,500	1700	85
Philadelphia	3,300	1500	65
St. Louis	2,300	1200	76
Los Angeles	2,600	1700	56
Cincinnati	1,900	950	81

Table 28. *Lithographic union membership, 1893–1960: membership on December 31 of each year.*

Year	Total	Journeymen	Appren-tices	New York Local 1	Chicago Local 3	San Francisco Local 17	No. of locals
			LIP&BA membership only				
1893	1,048 (mid-year)		none				15
1895 (April)	1,165		none	514	132	estab. June, 1895.	15
1897 (April)	1,512		none	696	126		16
1901 (April)	1,994		none	842	137		20
1904 (April)	2,873	2,873	none	1,174	126	86	23
1905	2,753	2,753	none	1,200	n.d.	89	24
1906	2,412	2,412	none	1,047	220	defunct	21
1907	1,492	1,492	none	806	128	defunct	19
1908	1,409	1,409	none	771	136	defunct	18
1909	1,601	1,601	none	779	165	defunct	19
1910	2,046	2,046	none	849	233	72	22
1911	2,368	2,368	none	915	255	105	26
1912	2,604	2,384	220	950	271	116	27
1913	2,792	2,562	230	1,017	292	109	30
1914	3,085	2,770	315	1,085	315	98	33
		Amalgamated Lithographers of America (ALA)					
1915	4,199	3,738	461	1,527	520	98	34
1916	4,574	4,096	478	1,632	609	93	34
1917	4,958	4,472	486	1,792	657	96	36
1918	5,686	5,195	491	2,323	693	119	40
1919	6,879	6,151	728	2,703	856	196	40
1920	7,619	6,868	751	2,928	833	249	46
1921	7,351	6,710	641	2,811	821	259	46
1922	5,754	5,343	411	2,374	672	217	46
1923	5,387	5,064	323	2,130	620	168	45
1924	5,322	5,006	316	2,096	596	158	45
1925	5,466	5,133	333	2,175	625	157	45
1926	6,027	5,553	474	2,372	689	175	45
1927	5,906	5,386	520	2,398	721	172	45
1928	5,624	5,172	452	2,383	694	185	46
1929	5,626	5,182	444	2,434	715	169	47
1930	5,717	5,224	493	2,460	753	149	47
1931	5,518	5,095	423	2,394	714	163	47

Table 28. (*Continued*)

Year	Total	Journeymen	Appren-tices	New York Local 1	Chicago Local 3	San Francisco Local 17	No. of locals
1932	5,207	8,864	343	2,313	679	155	47
1933	5,883	5,368	515	2,391	832	229	47
1934	6,694	5,760	934	2,550	990	351	48
1935	6,877	5,835	1,042	2,655	1,071	396	48
1936	7,200	6,019	1,181	2,827	1,073	443	51
1937	12,232	7,299	1,965	4,656	1,220	1,012	50
1938	12,033	7,464	1,922	4,383	1,257	953	51
1939	12,421	7,639	1,906	4,653	1,272	934	51
1940	12,849	8,022	2,017	4,610	1,505	929	52
1941	14,149	8,559	2,257	4,890	1,839	945	56
1942	13,250	8,582	1,885	4,706	1,869	755	56
1943	13,788	9,041	1,831	4,600	1,837	719	58
1944	14,635	9,498	1,867	4,976	1,914	720	59
1945	16,429	10,183	2,258	5,784	2,145	821	60
1946	20,718	11,464	3,806	6,888	2,704	1,105	65
1947	23,028	12,342	4,517	7,223	3,111	1,369	69
1948	23,421	12,806	4,642	6,908	3,340	1,455	72
1949	24,500	13,387	5,047	6,891	3,596	1,499	73
1950	25,232	14,334	4,913	6,824	3,727	1,535	74
1951	26,579	15,321	5,021	6,866	3,806	1,692	77
1952	27,768	16,549	5,032	6,997	3,929	1,748	78
1953	29,150	17,602	5,170	6,972	4,050	1,856	80
1954	29,235	19,152	4,864	6,900	4,157	1,941	81
1955	30,453	20,233	4,827	6,966	4,307	2,030	82
1956	32,340	21,266	5,248	7,340	4,563	2,098	83
1957	34,133	22,562	5,551	7,558	4,730	2,151	85
1958	35,087	23,450	5,674	7,627	4,853	2,145	89
1959	35,995	24,609	5,495	7,699	4,919	2,161	93
1960	37,099	25,716	5,652	7,868	4,987	2,169	93
1961	37,959	26,280	5,822	8,058	5,053	2,218	95

Sources: 1893: LIP&BA *Proceedings* (Boston, June 4, 1895), p. 14; 1895–97: LIP&BA *Proceedings* (Cincinnati, July 12, 1897), p. 23; 1901: LIP&BA *Proceedings* (New York, July 8, 1901), no paging; 1904: LIP&BA *Proceedings* (Philadelphia, July 11, 1904), p. 48; 1905–61: ALA Membership Statistics, internal record of the union.

Bibliography

Source material has come primarily from the following libraries, associations, and individuals: Dewey Library, Department of Economics and Social Science, Massachusetts Institute of Technology, Cambridge, Massachusetts; Baker Library, School of Business, Harvard University, Cambridge, Massachusetts; The Johns Hopkins University Library, The Johns Hopkins University, Baltimore, Maryland: New York Public Library, New York; Library of Congress, Washington, D.C.; Graphic Arts Institute of New England, 146 Summer Street, Boston, Massachusetts, Howard Patterson, secretary-manager; Printing Industry of America, Inc., 5728 Connecticut Avenue, Washington, D.C., George Mattson, lithographic labor relations advisor; Amalgamated Lithographers of America, 143 West 51st Street, New York, Donald Stone, secretary-treasurer; Niagara Lithograph Company, 1050 Niagara Street, Buffalo, New York, Carl Reed, executive vice-president; Arthur Willis, past president, Local 3, ALA, Boston, Massachusetts; Martin Grayson, past ALA international vice-president, now regional manager, Printing Developments, Inc., 540 South Michigan Avenue, Chicago, Illinois; Benjamin Robinson, general counsel, ALA, 100 East 42nd Street, New York; William Winship, president, Brett Lithographing Co., 47-07 Pearson Place, Long Island, New York.

There are five basic types of union records: (1) convention *proceedings,* (2) the *Lithographers Journal,* (3) *minutes* of special meetings, (4) internal records and documents of the union, (5) collective agreements.

PROCEEDINGS OF CONVENTIONS

Convention *proceedings* of the Lithographers' International Protective and Beneficial Association prior to 1895 and those for 1907 and 1911 are not available. *Proceedings* for 1895, 1897, 1901, 1904, and 1906 are in the Johns Hopkins Library; *proceedings* for 1904 and 1906 can also be found at ALA headquarters.

Convention transcripts of the Amalgamated Lithographers of America, 1917, 1919, 1923, 1925, 1927, 1930, 1939, 1946, 1947, 1949, 1951, 1953, 1955, 1957, and printed (and edited) *proceedings* from 1930 forward are in the files of the ALA; for 1951, 1953, and 1955, they can be found in the Dewey Library, M.I.T.

LITHOGRAPHERS JOURNAL

Complete bound volumes, 1915 to date, are in the New York Public Library; scattered early issues from 1921 to date can be found at ALA headquarters; complete copies from 1953 to date are in the Dewey Library, M.I.T. (Scattered issues of the *Bulletin of the LIP&BA,* which was irregularly published, are available at the Johns Hopkins University library.)

MINUTES OF SPECIAL MEETINGS

The *minutes* of international council meetings from 1915 to 1950 and those of the 1913 Buffalo amalgamation meeting are in the possession of Martin Grayson of Printing Developments, Inc.; a transcript of the 1947 special meeting on negotiations and the *proceedings* of the 1958 Cleveland Policy Conference are in the offices of the ALA.

INTERNAL RECORDS AND DOCUMENTS OF THE UNION

General letters to all local presidents and international councillors and scattered LIP&BA and ALA letters prior to 1921 belong to Martin Grayson of Printing Developments, Inc.; ALA letters, 1921 to date, are in the ALA files (an incomplete collection). Miscellaneous material and correspondence relating to the AFL jurisdictional award are in Martin Grayson's possession; relating to the NRA period, in ALA headquarters; relating to the Joint Lithographic Advisory Council, and action before the NLRA, in the possession of Benjamin Robinson of the ALA.

William Winship of Brett Lithographing Co. has *wage surveys* for 1943; the ALA has the surveys for 1947, 1952, and 1956 (recent material unavailable). The ALA also has records of the local bylaws (current and early scattered ones), the *Constitution,* and membership statistics by locals and class (journeyman, apprentice, general worker).

COLLECTIVE AGREEMENTS

The ALA has the current and recent agreements; Arthur Willis of the ALA has the Boston agreements, 1939 to date. The New York agreements, 1939–1946, are in Martin Grayson's possession.

Employer records are scattered, and many, notably of the Lithographers' National Association, have not been available. As in the case of union material, the listing below is to specify location of material, rather than its content.

LITHOGRAPHERS' NATIONAL ASSOCIATION RECORDS

Convention *Proceedings,* 1889 (NLA) are held by Carl Reed of Niagara Lithograph. From 1907 forward NAEL (LNA from 1927) annual conventions are reported in varying detail in the *National Lithographer* (for location see below). Convention *Proceedings* for 1937–1953 and 1955 are in Baker Library, Harvard University.

NAEL *shop rules* for 1907 are in the possession of Martin Grayson; *shop rules* for 1916 can be found at Baker Library, Harvard University.

The *National Lithographer* (a trade journal and not a publication of the LNA) is available from 1907 to date in the Library of Congress, Washington, D.C. Earlier issues are not obtainable.

INDUSTRY ECONOMICS

The Printing Industry of America's *Ratios for Printing Management* from 1922 to date are available at the Graphic Arts Institute of New England; from 1926 to date, they are to be found in Baker Library. PIA's *Results of Survey of Sheet-fed Offset Press Manning* (November 21, 1956) can be located at the Printing Industry of America, Inc. Lithographers' National Association/National Association of Photo-Lithographers, *Statistics Relating to Expansion of Lithographic Press Capacity* (May 1952) and statistics on postwar offset press sales, by press size of geographic area are also in Printing Industry of America, Inc.

DOCUMENTS RELATING TO NEGOTIATIONS

The statistical presentation in the 1958 New York negotiations is held by William Winship of Brett Lithographing Co. The *minutes* of the 1958 Boston negotiations and the transcript of discussion for the negotiations (joint union-management meeting held after signing of contract) can be found in the Boston Graphic Arts Institute.

The member manuals on *Prevailing wage information* of the LNA, NAPL, PIA are available. (The NAPL manual has more than adequate coverage and is the easiest to use.)

GOVERNMENT DOCUMENTS

U.S. Department of Labor, Bureau of Labor Statistics, *Earnings and Hours in Book and Job Printing: January 1942*, Bulletin no. 726.

——— *Union Wages and Hours: Printing Industry*, published once each year in the bulletin series, variously titled prior to 1948.

——— Statistical section in each *Monthly Labor Review* carries all BLS series of hours, earnings, and employment.

——— Wage and Hour Division, *Economic Factors Bearing on Minimum Wages in the Printing and Publishing and Allied Graphic Arts Industry*, September 1942.

U. S. Department of Commerce, Bureau of the Census, *Census of Manufactures, 1910*, 1 vol.; 1920, 1 vol.; 1929, 1 vol.; 1939, 4 vols.; 1947, 3 vols.; 1954, 3 vols.; 1958, Industry Reports.

——— *Annual Survey of Manufactures, 1955, 1956, 1957.*

——— *Biennial Census of Manufactures, 1921–1937* (excluding 1929).

BOOKS

Arbitration a Success! History of the Lithographers' Strike (New York: no publisher, 1896). The book was prepared by the International Lithographic Artists' and Engravers' Association, and is primarily a transcript of the arbitration hearings. 171 pp.

Baker, Elizabeth F., *Printers and Technology* (New York: Columbia University Press, 1957), 545 pp.

Brown, Emily C., *Book and Job Printing in Chicago* (Chicago: University of Chicago Press, 1931), 363 pp.

Dulles, Foster R., *Labor in America* (New York: Thomas Y. Crowell Company, 1949), 402 pp.

Dunlop, John T., *Industrial Relations Systems* (New York: Henry Holt and Company, 1958), 399 pp.

Halbmeier, Carl, *Senefelder and the History of Lithography* (New York: Senefelder Publishing Company [Carl Halbmeier], 1926), 216 pp.

Hoagland, Henry E., *Collective Bargaining in the Lithographic Industry* (New York: Columbia University Press, 1917), 130 pp.

Lithographers Manual, 2 vols. (New York: Waltwin Publishing Company, 1958), chapters paged individually.

Senefelder, Alois, *The Invention of Lithography*, translated from the original German by J. W. Muller (New York: Fuchs and Lang Manufacturing Company [a lithographic supply house], 1911), 229 pp.

ARTICLES

Dunlop, John T., "Structural Changes in the American Labor Movement," *Monthly Labor Review*, 80.2: 146–150 (February 1957).

Earle, S. Edwin, "The Lithographers' International Protective and Beneficial Association of the United States and Canada," *Journal of Political Economy*, 19.10: 866–83 (December 1911).

Jackson, John C., "Small Offset is Big Business," *Modern Lithography*, 54–56 (September 1958).

Thompson, Eldon H., "Methods for Converting Letterpress Forms to Offset," *The Inland Printer*, 183.3: 59–60 (December 1956).

Wilkinson, James S., "Summary of Small Offset Duplicators and Presses," *Modern Lithography*, 34–37, 149–150 (November 1956).

Notes

CHAPTER I. The System

1. Alois Senefelder, *The Invention of Lithography* (New York, 1911), p. 15.

2. John T. Dunlop, *Industrial Relations Systems* (New York, 1958).

3. Except where noted, the material in this section is drawn from Appendices A and B and the sources cited in those appendices.

4. Among them, Schmidt Litho in Milwaukee; McCandlish Litho in Philadelphia; Sackett and Wilhelms and Brett Litho in New York. I am indebted to Maurice Saunders, William Winship, and A. H. Wilhelm for this information.

5. Walter E. Soderstrom, "The History of Lithography," Lithographers Manual, I. 1:6–1:7 (New York, 1958).

6. *The Census of Manufactures* classified sixty-two firms as "lithographers" in the Boston Metropolitan area in 1954; 3000 in the country at large.

7. This figure and the one following for total employment in metal decorating are based on odd bits of information gathered from interviews with two members of the National Metal Decorators' Association and with President Spohnholtz of Local 4, Chicago. President Swayduck of Local 1, New York, was not able to provide information on this point. The Census dropped metal lithography as a classification in 1954, and no reliable figures are available. Since the receipts in 1947 were less than .5 per cent of the total (*1947 Census of Manufactures,* p. 373), it is probably not a serious omission. Metal lithographers not in captive plants are equally scarce in the ALA. Local 4, with over 600 metal lithographers, has 95 per cent of them in captive plants.

8. Interview, November 1958. Further information comes from a paper lithographer: "At no time were the metal decorators in the metropolitan area, in the Metropolitan, even though they were invited. At one time they asked to have an observer sit in at our negotiations, which we allowed, and then they negotiated a contract on their own, or at least, tried to. We thought this was unfair to the Metropolitan members and asked them to join with us so that we would have more strength and they would bear the proportionate cost of the negotiations, but they could never see fit to do this, and we have never had any more dealings with them. So far as I know, the metal decorator plants, whether they be captive or otherwise, are supposed to have the same contract as the paper plants, in fact, our contract stipulates that if the Amalgamated negotiates any better wages or working conditions in the area under the jurisdiction of Local 1, we automatically receive such wages and working conditions in our contract." (Letter to author, March 9, 1959.)

9. Speech of Richard J. Walters to Package Design Council, October 19, 1958. Other estimates, as in the case of Walters' speech based on statements by press manufacturers (not the same ones), place the percentage of large press sales going to the folding paper box industry at 20 per cent in 1958, sharply up from around 5 per cent in preceding years. (Letter from George Mattson to author, November 24, 1958.)

10. ALA Brief in Sutherland Paper Co. Case, 7-RC-3793, 1958, p. 3.

11. *See* in particular the Sutherland Paper Co. Case, 7-RC-3793, 1958.

12. Speech of Richard J. Walters, cited above.

13. Both the IPP&AU and the ALA have recognized this problem but have chosen widely different methods of dealing with it.

14. *National Lithographer,* October 1909, p. 7.

15. George Strebel, Printing Industries Association of Western New York (letter to the author, June 1960).

16. Harris-Seybold and American Type Founders provided the information covering the years 1946–1948. Small presses were defined as larger than offset duplicating machines but excluding two-color and large single-color presses. For Harris Seybold, nothing larger than the 17″ x 22″ was included. The number of presses sold was confidential, given only as "considerably in excess of a thousand." Thirty-three per cent of sales were to combination plants, and presumably the 6 per cent-plus that is unaccounted for was sales to captivé plants. "Combination plant" is undefined; it probably means any plant with letterpress equipment, plus at least one offset press, 17″ x 22″ or larger. *Inland Printer,* September 1948.

17. *Inland & American Printer & Lithographer,* August 1959, pp. 72-74. The figures are based on answers to a survey from 24 per cent of the magazine's subscribing plants, subscribers being described as including "virtually every large and medium-size plant and most progressive small ones." The number is not given.

18. Interview with Edward Gruen, Chief of Printing and Publishing Industry Division, Bureau of the Census, Novmeber 1958.

19. Henry C. Latimer, Metropolitan Lithographers Association, cited in his article in *75 Years of Lithography,* Supplement to *Lithographers' Journal,* September 1957, p. 22. See *1958 Census of Manufactures,* pp. 273–274 for Census sources.

CHAPTER II. Key Actors in the System

1. *Poor's Register of Directors and Executives, 1958* lists some 240 incorporated firms producing lithography. Not all of them conform to the Census definition since folding paper box plants and metal decorators are included. Of a random sample of 50 drawn from this group, in 29 cases the name of the senior officer or officers was the same as that of the firm, or at least two of the top three officers had the same last name. Presumably, the number of firms that are misclassified as family-owned and -operated by these standards are compensated for by the number missed which are owned and operated by two or three families.

2. Estimate of Howard Patterson, former secretary-manager of the Graphic Arts Institute of New England.

3. Professor Kenneth Scheid of Carnegie Institute of Technology, also a student of this subject, has indicated to me that his data (not yet published) led to a less extreme conclusion.

4. Material drawn from interview notes, October 1957. Not an exact quotation of the interviewee.

5. Interview with trade association executives, November 1958.

6. Interview with trade association executive, November 1958.

7. With the addition of the Label Manufacturers Association in 1959, the LNA changed its name to the "Lithographer *and Printers* National Association." To avoid confusion, the earlier title is used throughout the study.

8. *National Lithographer,* January 1927, p. 31. *See also* Paul A. Heideke, "Early Days of NAPL," *Modern Lithography,* September 1956, p. 91.

9. Printing Industry of America, *Annual Convention Program,* 1952, p. 26.

10. *LNA Directory,* no date, p. 1.

11. Elizabeth Baker, *Printers and Technology* (New York, 1957), pp. 380–84.

12. Over half the LNA's members in 1958 were also members of PIA. An even greater overlap is likely between NAPL and PIA, but NAPL does not publish the names of member firms, and this remains a conjecture.

13. ALA *Proceedings,* Atlantic City, 1939, officers' reports, p. 20.

14. Harold A. Logan, *Trade Unions in Canada* (Toronto, 1948), pp. 5-7.

15. Stuart M. Jamieson, *Industrial Relations in Canada* (Ithaca, 1957), p. 6.

16. ALA *Proceedings,* Atlantic City, 1939, officers' reports, pp. 20, 34.

17. ALA *Proceedings,* Dallas, 1951, officers' reports, p. 55; *Proceedings,* Chicago, 1957, officers' reports, p. 27.

18. This paragraph is based on interviews with George Luke, President, Local 2, Buffalo; Carl Reed, Niagara Lithograph, Buffalo; and George Streble Executive Vice-President, Printing Industry of Western New York, Buffalo, July 1958.

19. The material in the following three paragraphs is based on interviews with Henry Latimer, Executive Director of MLA; William Winship, President, MLA, and President of Brett Litho; Don Taylor, Secretary, Printers' League Section, New York Employing Printers Association; and Edward Swayduck, President, Local 1, ALA.

20. Material in this paragraph gained during contact with the 1958 negotiations in Boston.

21. Interview with A. T. Howard, January 1959.

22. Interview notes, July 1958; July 1960.

23. Interview, September 1958. The quotation is from a private transcript.

24. Interview with James Armitage, Inland Press, Chicago.

25. Membership data are taken exclusively from *ALA Membership Statistics, 1905–1959.* Unless otherwise noted, they are for December 31, 1959.

26. The Census of Manufactures and Bureau of Labor Statistics gather data for industries, not occupations. The occupational data of the decennial population census puts lithographic pressmen with other printing pressmen, and the preparatory classifications with photo-engravers in a "Photo-engraver and Lithographer" grouping. The method of matching these data with ALA statistics is described in Appendix B.

27. Total union membership, according to Bureau of Labor Statistics, *Monthly Labor Review* 83:1.2 (January 1960) was 14 million in 1950 and slightly over 17 million in 1958. The industry employment figure for 1950 is calculated by assuming a constant increase between 1947 and 1954, the two years bracketing 1950 in which complete censuses were taken for the Census of Manufactures. Annual surveys are too inaccurate to be useful.

28. The standard exception is because two councillors will almost certainly come from Local 1, New York.

29. ALA *Constitution* (1958 revision), article IX, section 6.

30. ALA *Proceedings of the National Policy Conference,* October 1958, p. 168.

31. Interview, November 1958.

32. The departments are art, engraving (usually not represented), camera, platemaking, stripping, press, feeder and floor help, general workers.

33. Boston, New York, Baltimore, Rochester, Buffalo, Toronto, Chicago.

34. Interview notes, July 1958.

CHAPTER III. The March of Technology

1. *American Pressman,* January 1960, pp. 26–27.

2. Recent developments, notably the du Pont photopolymer plate, may well eliminate this disadvantage to letterpress.

3. For example, *see* the description of the Kodak three-color process, in *Lithographers Manual*, II. 7:21, Victor Strauss, ed. (New York, 1958). The section is prepared by J. A. C. Yule, Research Associate, Applied Photography Division, Kodak Research Laboratories. The section carries a 1957 Eastman Kodak copyright.

4. ALA *Proceedings*, Chicago, 1957, p. 152. In 1957 the ALA believed that thirty scanners of the type developed by Printing Developments, a subsidiary of *Time*, Inc., could handle all the color work in the United States and Canada.

5. "Questions and Answers from the NAPL Technical Session," *Modern Lithography*, November 1956, pp. 38–39.

6. Excerpt from advertising brochure of Davidson Corp., subsidiary of Merganthaler Linotype (1958).

7. Excerpt from advertising brochure on nuArc Rapid Printer, nuArc Company, Inc. (1958).

8. There is much controversy about entry costs to the lithographic industry; the range given here is based on price lists (new and used equipment) plus $1000 installation costs. An American Type Founders' executive has estimated entry costs at $6500. John C. Jackson, "Small Offset Is Big Business," *Modern Lithography*, September 1958, p. 56.

9. ALA *Proceedings*, Boston, 1955, pp. 285–286.

10. This and several other transfer methods are discussed in Eldon H. Thompson, "Methods for Converting Letterpress Forms to Offset," *The Inland Printer* (December 1956), p. 59.

11. ALA *Proceedings*, Portland, Oregon, 1959, p. 63.

12. ALA *Proceedings*, Portland, Oregon, 1959, pp. 102–106. One exception was the rebuttal of Local 4 (Chicago) Vice-President George Gunderson.

13. ALA, *Prevailing Wage Survey*, December 1956. Internal record of the union.

14. ALA *Proceedings*, Portland, Oregon, p. 62; pp. 133, 135.

CHAPTER IV. The Structure of Markets

1. Interview with William Gruber, Boston, October 1957.

2. Interview with Douglas Reilly, Buck Printing, Boston, November 1957.

3. Interviews, Boston, October-November 1957.

4. This range draws on the statements of lithographers interviewed in the Boston area and the results of a survey reported in "If You Were to Start in Offset Again, What Would You Do?" *New England Printer and Lithographer*, August 1956, p. 41.

5. Professor Kenneth Scheid of the Carnegie Institute of Technology's School of Printing Management has suggested that the practice of setting up a small firm to handle commercial and short-runs of color lithography may grow as an alternative to profit plowback in the large firm.

6. Frank J. Turner, "B'guess and B'gosh . . . or Who's Kidding Whom about Costs," *New England Printer and Lithographer*, September 1958, pp. 64–65.

7. Interview with George Mattson, Printing Industry of America, November 1958.

8. Ed Larson, "Selling Slants," *New England Printer and Lithographer*, 21.5 (June 1958).

9. James W. Sheldon, "How Well Do You Estimate," *New England Printer and Lithographer*, January 1956, p. 85.

10. H. T. Koerner, "Notes on Lithography and the Costs of Its Production," *First Annual Report*, National Lithographers' Association, Buffalo, New York, 1890, p. 91.

11. During a discussion of the high cost of union operations in one of the ALA conventions during the 1920's, economizers secured the passage of such a resolution regarding printing purchases of the union. It was later modified on the somewhat embarrassing grounds that its enforcement would require the officers to buy all their printing from nonunion shops.

12. Interview, January 1959.

13. Statement to the author by a trade association executive, July 1960.

14. Interview with Douglas Reilly, Buck Printing, Boston, October 1957.

15. Henry C. Latimer, "History of the Lithographic Industry," *Lithographers Journal,* 75th anniversary edition, part II, September 1957, p. 21.

16. The slope of lines X, A, and B has been chosen arbitrarily in Diagram B. But if the price axis is given numerical values at each point, or even expressed as percentage changes from some norm, then the slope of line X would measure the imperfections existing in the market in the absence of the buyer-supplier bond, and the slopes and distance between A and B would measure the strength of that bond.

17. A copy of the Jiffy Estimator has been made available to me through the courtesy of Arthur Willis, ex-president of Local 3, ALA (Boston).

CHAPTER V. The Workforce

1. Both the Bureau of the Census and the Bureau of Labor Statistics use the Bureau of the Budget's *Standard Industrial Classification Manual* for industry definitions, and the recommended "employee" and "production worker" definitions.

2. John T. Dunlop, "The Task of Contemporary Wage Theory"; E. Robert Livernash, "The Internal Wage Structure," *New Concepts in Wage Determination,* edited by George W. Taylor and Frank C. Pierson (New York, 1957), pp. 117–172. For a discussion of job clusters, *see* pp. 129–30.

3. Sumner H. Slichter, *Union Policies and Industrial Management* (Washington, 1941), pp. 32–34.

4. There is reason to believe that the IPP&AU had more lithographic press feeders among its membership in 1917 than it had lithographic workers of all types at any time during the 1920's. The ALA *State of Association Report* in 1917 referred specifically to six locals in which the IPP&AU had feeder members. One of the effects of amalgamation of the lithographic unions was to make it possible for the ALA to force such men into their organization.

5. In part, this is based on an interview with Vice-President Rohan of the IPP&AU, November 1958.

6. Union Employers Section, PIA, *Results of Survey of Sheet-Fed Offset Press Manning* (November 21, 1956). Made available through the courtesy of William Reinhardt, Rand McNally, Chicago.

7. International President (then Vice-President) D'Andrade has estimated that the IPP&AU had 8000 offset pressmen and preparatory workers in 1957. Interview, April 1959.

8. Agreement between Contract Employers' Group (Lithographers' Division) and Local No. 3, ALA, September 22, 1958. Estimates of coverage are those of Arthur Willis, ex-president, Local 3, Boston.

9. The IPP&AU has fewer members qualified to move into such jobs, but a more important reason excludes them from consideration. The IPP&AU has had a long experience in following the printing press into whatever industry it might be employed. This experience has caused it to adopt a quite different tactic from the ALA — that

of accepting lesser working conditions for the same jobs when found outside the printing industry than within the industry. For this reason employers are less violently opposed to the IPP&AU than the ALA but could hardly expect the IPP&AU to urge skilled craftsmen to move to such employment.

10. *Constitution of the Amalgamated Lithographers of America,* as revised to 1958: article XII, section 3 of "Constitution and General Laws as They Apply to the Locals," p. B13; article VII, section 4, "International Constitution," p. 9.

11. The states are Kentucky, Alabama, Tennessee, Oklahoma, and Texas. Figures for the comparison are taken from the *1947* and *1958 Census of Manufactures,* and from internal records of the union. During the period 1947–1958, lithographic employment in the region rose by 225 per cent, union membership by 60 per cent.

12. ALA, *Minutes of Special Negotiating Meeting,* New York, May, 1947, p. 247.

13. Interview with President (then Vice-President) Turner, Local 12, Toronto (July 1958).

14. *Agreement* between the Metropolitan Lithographers Association, Inc. and Amalgamated Lithographers of America Local No. 1, May 1, 1958 to April 30, 1960. article 16, clauses a) and f).

CHAPTER VI. The System Before the Thirties

1. In 1915 Henry Hoagland did a careful study of collective bargaining in the lithographic industry, concentrating on the period 1886–1906. It is a scholarly, well-documented study and is used extensively in the early part of this chapter. Unless otherwise noted, factual material has been drawn from his book, and only direct citations from him are individually noted. Henry E. Hoagland, *Collective Bargaining in the Lithographic Industry* (New York, 1917).

2. Philip S. Foner, *History of the Labor Movement in the United States* (New York, 1947), p. 236.

3. An original copy of the constitution of this association is in the possession of Walter Soderstrom, executive vice-president of the National Association of Photo-Lithographers (NAPL). His contribution of this and other information is gratefully acknowledged.

4. Hoagland, *Collective Bargaining,* p. 17.

5. Union sources are Lewis Cass Gandy, "The Story of American Lithographers," *Lithographers Journal,* March 1940 (supplement, independent paging); Martin Grayson, *Amalgamated Lithographers of America: Some Notes on Its History,* a pamphlet prepared originally for the union's mountain regional conference of September 1954.

6. National Lithographers' Association, *First Annual Report,* 1890. I am indebted to Carl Reed of Buffalo for the use of this book.

7. Hoagland indicates that this broadened jurisdiction predates the strike, quite probably basing his statement on the charter granted by the Knights, which the author has not seen.

8. For a brief mention of these unions *see* the LIP&BA *Proceedings,* Philadelphia, 1904, pp. 33–35, 100–103.

9. National Lithographers' Association (NLA), *First Annual Report,* p. 24. See also Hoagland, *Collective Bargaining,* p. 18.

10. NLA, *First Annual Report,* pp. 21–22, 27, address of President Julius Bien.

11. Hoagland, *Collective Bargaining,* pp. 20, 83.

12. LIP&BA *Proceedings,* Boston, 1895, p. 10. *Proceedings* of the Third Biennial Convention of the LIP&BA, p. 10, cited in Hoagland, *Collective Bargaining,* p. 22.

13. Material in this paragraph is drawn from Hoagland, *Collective Bargaining,* pp. 23–28, and from *Arbitration a Success! History of the Lithographers' Strike* (New York, 1896). The latter book was printed by the union for distribution to its membership. I am indebted to Carl Reed of Buffalo for the use of his copy.

14. No figures are given for the feeders; the estimate is based on a discussion appearing on p. 14 of the 1901 convention *proceedings,* New York. Although the LIP&BA had among its members all crafts but feeders, it actively organized only transferrers, pressmen, and provers (pressmen who make only proofs), and refused admittance to additional stone grinders.

15. Buffalo and Rochester still had 1¼, Louisville 1⅓, and Baltimore 1⁵⁄₁₂ overtime rates. Other crafts had not yet secured even this degree of uniformity.

16. Report to the Convention, LIP&BA *Proceedings,* New York, 1901, p. 5 (separate paging).

17. *See* report of General President Keough in LIP&BA *Proceedings,* Cincinnati, 1897, p. 8; also Hoagland, *Collective Bargaining,* p. 30.

18. Among the leading employers were the following (all but two were also charter members of the NLA, when it was formed in 1889): Julius Bien (Julius Bien & Co.); A. Wilhelms (Sackett & Wilhelms); Olin D. Gray (Gray Litho Co.); H. P. Bailey (Trautmann, Bailey & Blampey); Charles W. Frazier (Brett Litho Co.); William Ottman (J. Ottman Litho Co.); R. M. Donaldson (American Litho Co.).

19. *Lithographers' Bulletin,* July 1, 1903, p. 4. (irregular publication). Concern was expressed about the rule in the 1901, 1904, and 1906 conventions of the LIP&BA.

20. LIP&BA *Proceedings,* Buffalo, January–February 1906, pp. 146–147.

21. Hoagland, *Collective Bargaining,* p. 55.

22. LIP&BA *Proceedings,* Philadelphia, 1904, p. 100.

23. The eight-hour day was a vital issue throughout the printing trades at this time. The initial strikes to enforce this demand in commercial shops began in mid–1905, although the shorter hours were not generally introduced until 1907. Emily Clark Brown's *Book and Job Printing in Chicago* (Chicago, 1931) gives an excellent summary of the struggle.

24. From its inception until well into the 1930's the NAEL was considered *the* association in the lithographic industry by the public, by the employers, and by the union. The material in the paragraph above is drawn from Hoagland's study, as is the major portion of the appendix.

25. NAEL general letter to members, cited in Hoagland, *Collective Bargaining,* p. 92.

26. Estimate of the author, derived from statements in Hoagland, *Collective Bargaining,* p. 103; *National Lithographer,* May 1907, p. 12; and LIP&BA *Proceedings,* Buffalo, 1906, pp. 18–20, 213–214.

27. NAEL General Letter No. 104, cited in Hoagland, p. 96.

28. NAEL General Letter No. 20, August 23, 1906, quoted in Hoagland, p. 96.

29. *Lithographers Journal,* November 1916, p. 9n.

30. The estimate is based on statistics given in Hoagland, p. 103.

31. NAEL General Letter No. 640, cited in Hoagland, p. 111.

32. LIP&BA *Proceedings,* Philadelphia, 1904, p. 2. This would include the George Harris Co. and Ketterlinus Litho, two of the very large lithographic firms in the country.

33. Hoagland records that in 1915 there were 121 member firms with 845 to 850 presses, having about 5000 "employees." This cannot mean total employees, or the average employment per firm would be only 41 workers. This is absurd since, ac-

cording to the Census, average employment for the whole industry was 45 per firm, and the association drew its members from the large employers. If we assume that the figures given by Hoagland refer to the lithographic crafts and exclude all others, employment in association firms works out to be about 50 per cent of that for the industry.

34. Statement by Frank Gehring in *Lithographers Journal*, I. 1:3 (June 1915).

35. To distinguish this meeting from other important Buffalo conferences, it is usually referred to as the one held in Peterson's Hall. Frank J. Peterson was a power in the LIP&BA during this period, but his lasting claim to fame stems from his defection from the union in the 1922 strike, when he led the Buffalo local out of the ALA and made an important contribution to the ALA's defeat.

36. Actually, of the International Union of Lithographic Workmen, the title which the league had adopted in 1912.

37. Comparable to the general executive board of the LIP&BA. The name was changed to the international council in the 1917 convention.

38. ALA *Proceedings*, I:90 (Chicago, 1919).

39. ALA Proceedings, Cincinnati, 1917, pp. 324–328.

40. Letter from Secretary-Treasurer O'Connor to all locals, September 27, 1918.

41. *Lithographers Journal*, November 1915, p. 31.

42. Letter from President Bock to P. D. Oviatt, September 18, 1917. Curiously, Charles F. Traung, who was vice-president and a member of the general executive board of the LIP&BA in the earlier period, was now on the board of directors of the NAEL.

43. *National Lithographer*, June 1911, p. 18.

44. Letter to all members, August 25, 1919, gives the text of the agreement.

45. *National Lithographer,* September 1919, p. 31.

46. "Cost of Living in the United States," *Monthly Labor Review,* February 1923, p. 357. The index is based on data gathered from thirty-two large cities. No figure for June 1919 is available.

47. *Lithographers Journal*, December 1919, p. 249; January 1920, p. 287; February 1920, p. 329; March 1920, p. 363.

48. Lithographic increases compiled from ALA *Proceedings* I:100–193 (Chicago, 1919); letterpress increases compiled from Emily Clark Brown, *Book and Job Printing*, pp. 157–171.

49. ALA *Proceedings*, I:158 (Chicago, 1919).

50. Letter to all members, June 1, 1920, gives the text of the agreement.

51. *National Lithographer*, March 1921, p. 39.

52. *Lithographers Journal*, November 1920, p. 193; January 1921, p. 264; ALA *Proceedings*, III:1222 (Cleveland, 1923).

53. *Monthly Labor Review*, February 1923, p. 357. Figures are based on data gathered from thirty-two major cities.

54. *National Lithographer*, April 1921, p. 40; September 1921, pp. 27–30.

55. No statistics are available to measure the intensity of unemployment during this period. Unemployment is talked about in the trade journal and union journal, but no useful figures are given. Census figures make it possible to estimate that actual 1921 employment in the industry was at least 10 per cent less than what comparable data for 1914, 1919, 1923, and 1925 would lead one to expect, but they throw no light on the trend of unemployment during the year.

56. By the terms of the 1919 and 1920 settlements, a July 1919 wage of $20 in association shops would have increased to $30 in the latter part of 1920. Average

weekly wage figures comprise wages divided by wage earners, as reported in the *Census of Manufacturers.*

57. Letter addressed to all local presidents, October 8, 1921, signed by the nine-man union committee.

58. Members' general letter N–1458, September 14, 1928.

59. Letter from President Bock to all local presidents, April 19, 1927.

60. *National Lithographer,* December 1921, p. 30, reports the employer conference at which the decision was reached.

61. *National Lithographer,* January 1922, p. 33; February 1922, p. 26; November 1922, p. 45. Letters from Secretary-Treasurer O'Connor to local presidents, January 5, 11, 19, and 27, 1922. See also *ALA Proceedings,* I:280 (Cleveland, 1923).

62. Members' general letter 1381–A, February 23, 1923.

63. Letter from NAEL secretary to an NAEL director, July 10, 1924.

64. *National Lithographer,* November 1922, p. 40.

65. Members' general letter no. 1405–B, August 12, 1924; No. 1429–A, December 11, 1926; No. 1386–F, June 26, 1923; No. 1426, September 15, 1926; No. 1439, August 25, 1927.

66. LNA director to LNA counsel, July 21, 1924.

67. Letters, December 4, 1926, and December 31, 1926.

68. The head of Forbes Lithograph in Boston was worried, and properly so, when he heard that Fred Rose was spending several days organizing in Boston. W. S. Forbes to Earl Macoy, April 5, 1927.

69. Robert Bruck, international vice-president, *Report to the International Council,* January 1, 1929, pp. 1–3.

70. Letter from LNA President John Omwake to directors, August 25, 1926. The percentage cited by Omwake was 17.8.

71. Members' general letter 1486–C, May 21, 1930. Membership was about 100 firms during this period.

72. Members' general letter 1489–C, July 2, 1930.

73. Letter to another LNA director, May 5, 1927.

74. Taken from a joint statement issued by employer and union representatives at the close of a December 10 conference, and approved by a special meeting of Local 1 on December 19. Reported in the *Lithographers Journal,* January 1928, p. 375.

CHAPTER VII. The System During the Thirties and the War Period

1. ALA *Preceedings of the Fifth Convention* II:734–735 (Toronto, 1927).

2. ALA *Proceedings* III:1122 (Cleveland, 1923).

3. ALA *Proceedings of the Fifth Convention* II:608–609 (Toronto, 1927).

4. Bureau of Labor Statistics, *Monthly Labor Review,* March issues of vols. 28, 30, 32 and 36. No monthly data available on the lithographic industry.

5. No statistics are available before 1932. The estimate is from the *minutes* of the 1930 convention. The 1932 figures are from *International President's Report,* p. 6, contained in *Proceedings,* Atlantic City, 1939.

6. Foster Rhea Dulles, *Labor in America* (New York, 1949), p. 261.

7. A former officer of the LNA challenges the accuracy of this statement. I do not mean that the *National Lithographer* was an official organ of the LNA. I do mean that it had no editorial position on labor questions which could be distinguished from that of the LNA, up to and including the proposition that firms were doing a disservice to the industry if they did not join the industry association. It is of interest in this

connection that the *National Lithographer* was once officially designated as the journal for printing all LIP&BA notices. Its founder, Richard C. Norris, was also a founder of the LIP&BA. Upon his death his son became editor and shortly sold the *National Lithographer* to Warren C. Browne. It was under the latter's editorship that a gradual shift in sympathies occurred.

8. *National Lithographer,* January 1927, p. 31.

9. Material in this section has been taken from the following sources: National Archives, *Preliminary Inventories: Number 44, Records of the National Recovery Administration* (Washington, 1952), pp. 1–5; National Recovery Administration, Division of Review, *Classification of Approved Codes in Industry Groups* (Work materials no. 12), pp. 1–22 (mimeo.); Lewis Mayer, *A Handbook of NRA,* 2nd ed. (New York, 1934), pp. vii–ix, 439–442.

10. President Roosevelt's *Message to Congress,* January 3, 1934.

11. Cold storage door, ring traveler, animal soft hair, milk filtering material, etc., horse hair dressing, brattice cloth, blackboard and blackboard eraser, steel joists. Covered employment only given to the nearest 100. *Classification of Approved Codes,* pp. 1–22.

12. Members' general letter 1549–N, May 5, 1933.

13. Letter from President Kennedy to members of the international council, August 12, 1933.

14. *The Code of Fair Competition for the Graphic Arts Industries,* Code 287, approved February 17, 1934, actually was made up of a graphic arts coordinating committee, four compliance boards — one each for the letterpress, lithographic, gravure, and printing service industries — fifteen code authorities, seventeen national product groups, three labor boards, together with regional compliance boards, code authorities and labor boards to handle local administration of the codes. Apparently only six of the code authorities were active, and only the lithographic labor board was actually set up. It is not known how many of the product groups were operative.

15. Letter from President Kennedy to all officers and members, June 23, 1936.

16. *Lithographers Journal,* June 1935, p. 34.

17. Bureau of Labor Statistics, *Trade Agreements: 1925,* bulletin no. 419 (September 1, 1926), p. 116.

18. *Report* of the deputy administrator to the administrator, undated, copy on file in ALA internal records.

19. Letter from President Kennedy to members ofthe international council, May 1, 1934.

20. Letter from President Kennedy to members of the international council, May 14, 1935.

21. President Kennedy's Report to the 1939 convention, ALA *Proceedings,* Atlantic City, 1939, official reports and resolutions, p. 1.

22. President Roosevelt's statement on signing the NRA, June 16, 1933, cited in full in Lewis Mayer, *A Handbook of NRA* (New York, 1934), pp. 80–85.

23. LNA *Proceedings,* titled *Market Opportunities for Lithographers* (1937), secretary's annual report, p. 77.

24. Vice-President Fred Rose in *Lithographers Journal,* April 1934, p. 11.

25. Several drafts of the agreement were printed, and copies are available in both association and ALA files. The one used here was the one agreed to on June 22, 1937, to go into effect June 29.

26. ALA *Proceedings,* Atlantic City, 1939, p. 14.

27. Letter from President Kennedy to all officers and members, December 13, 1938.

28. *National Lithographer,* October 1937, p. 13.

29. The 1923 figures based on State of Association Reports, ALA *Proceedings,* II:659–797 (Cleveland, 1923). The 1947 figures are based on 1947 *National Wage Survey,* internal record of the union. Weighted averages were not used to calculate the percentages.

30. Letter from John (Tricky) Forbes to Martin Liberatore (Local 3 Boston), September 30, 1937. See also Benjamin Robinson, *Memorandum to International Council,* March 6, 1944.

31. Interview with Local 11 (Rochester) President Hilsdorf, July 1958.

32. *Lithographers Journal,* January 1947, p. 636.

33. Letter from Paul Ocken, vice-president and general manager of Graphic Arts Industry, Inc., to Clarence Lofquist, reprsentative, IPP&AU, May 14, 1945. The complete letter is three pages long and contains similar illustrations.

34. Benjamin Robinson, *Memorandum* concerning preparation for September 15, 1944 meeting of JLAC, probably attached to a September 5 letter to President Riehl.

35. Bureau of Labor Statistics, *Trade Agreements: 1927,* bulletin no. 468 (December 1928), pp. 138–139.

CHAPTER VIII. The System in the Postwar Period

1. ALA *Proceedings,* Colorado Springs, 1946, p. 304.

2. Unless otherwise noted, material in this section is drawn from NLRB, in the Matter of Albert Love Enterprises (*Foote and Davies*) 66:416–452 (March 8, 1946), *Decisions and Orders of the National Labor Relations Board; In the Matter of Pacific Press, Inc.,* 66:458–464 (March 8, 1946).

3. A discussion of the AFL–CIO and NLRB attitudes toward jurisdiction appears in John T. Dunlop, "Structural Changes in the American Labor Movement," *Monthly Labor Review,* 80:146–150 (February 1957).

4. NLRB, *Foote and Davies,* pp. 417, 423, 438, 462–63.

5. ALA *Proceedings,* Colorado Springs, 1946, pp. 110–111.

6. ALA *Proceedings,* Boston, 1955, pp. 232–233.

7. Earle K. Shawe, *Brief on Behalf of the Employer, In the Matter of Sutherland Paper Company,* Case No. 7–RO–3717, September 15, 1958, pp. 40, 62–63. The brief has been made available through the courtesy of Mr. Shawe.

8. 107 NLRB 290. The decision itself explains the reasons for the shift in policy.

9. ALA *Proceedings,* St. Paul, 1949, p. 12 of September 20 session.

10. ALA *Proceedings,* Boston, 1955, officers' reports, p. 9.

11. ALA *Proceedings,* Chicago, 1957, pp. 77, 81–82.

12. *AFL–CIO News,* August 30, 1958, p. 3.

13. No title, probably done in September 1946. Courtesy of Benjamin Robinson.

14. ALA *Proceedings,* Colorado Springs, 1946, p. 98.

15. Blackburn was a member during the latter part of the JLAC's existence.

16. Letter from Local 1 President John Blackburn to Vice-President Robert Bruck, March 5, 1946. Floyd Maxwell had become secretary of the LNA in 1936.

17. ALA *Proceedings,* Colorado Springs, 1946, p. 193.

18. Benjamin Robinson, *Memo on CIO Jurisdictional Action,* January 3, 1952.

19. ALA *Proceedings,* Chicago, 1957. The whole speech is thirty-five pages long, and must have taken some three hours to deliver. The quotations are from pp. 79–80, 82–83, 85–86.

20. *New York Times,* August 26, 1958, p. 32n.

21. *See* speech by Councillor Donahue, ALA *Proceedings of the National Policy Conference,* pp. 148–164, 171.

22. For a discussion of this, *see* Baker, *Printers and Technology* (New York, 1957), pp. 130–140.

23. AFL *Proceedings,* St. Paul, 1918, p. 169.

24. Baker, *Printers and Technology,* p. 434.

25. *See* ALA *Proceedings,* Biloxi, Mississippi, 1947, officers' reports, p. 9; ALA *Proceedings,* St. Paul, 1949, officers' reports, p. 21.

26. Baker, *Printers and Technology,* p. 429n.

27. In 1951 the ITU is first referred to in ALA *Proceedings* as a potentially dangerous enemy.

28. Interview, November 1958.

29. Meeting, November 1958. Writer was present.

30. Quoted in full in union employers' section, Printing Industry of America, *Industrial Relations Letter,* November 23, 1960, p. 3.

31. ALA *Proceedings of the National Policy Conference,* Cleveland, 1958, pp. 154, 190.

CHAPTER IX. The Machinery of Bargaining

1. Lithographers' National Association, LNA *Members' Bulletin,* June 17, 1946, pp. 10, 12, 19. The bulletin includes a transcript of the O'Brien speech and the question period following it.

2. LNA *Members' Bulletin* 20420 (February 6, 1947).

3. ALA *Proceedings of the Special Negotiating Meeting,* New York, 1947, p. 23. This was a meeting of local leaders who were going into bargaining in the near future. Some 16 locals were represented. This will henceforth be cited as the ALA *1947 Negotiating Session.*

4. Letter to author, September 4, 1959.

5. The ALA *1947 Negotiating Session,* p. 20.

6. Letter to the author, July 1960.

7. The ALA *1947 Negotiating Session,* pp. 11–19, 24n, 189–190.

8. ALA *Proceedings,* Portland, 1959, pp. 214, 219, 221.

9. Letters to the author, May, July, 1960. There is some evidence that international representatives are taking a more active part in the 1960 negotiations.

10. ALA *Proceedings,* Portland, Oregon, 1959, p. 123. *See* the preceding speech of the vice-president and director of organizing, Jack Wallace, and the supporting comment of International Councillor Kenneth Brown, now international president.

11. Interviews in October 1957 and November 1958.

12. *Contract Discussion, Contract Employers Group and Local No. 3 Amalgamated Lithographers of America,* October 29, 1958, pp. 83–84. This will henceforth be cited as *Contract Discussion Transcript,* Boston, 1958.

13. The writer attended the Boston negotiations as a guest of the employers. Unless otherwise noted, material is drawn from notes made at the sessions.

14. Interview with James Armitage, August 1958.

15. Meyer S. Ryder, "Strategy in Collective Bargaining Negotiations," *Michigan Business Review,* November 1955.

16. ALA *1947 Negotiating Session,* pp. 15, 84–107, 275n.

17. *New York Times,* September 22, 1957, p. 1.

18. *New York Herald Tribune,* October 20, 1957, p. 27n.

19. ALA *Proceedings,* Chicago, 1957, pp. 218–219.

20. Interview, January 1958.

21. Bureau of National Affairs, 44 LRRM 2577; 45 LRRM 2710.

22. Metropolitan Lithographers Association meeting of November 20, 1958. After-dinner remarks by Benjamin Robinson and Edward Swayduck. The quote is taken from notes made at the meeting, and is not exact.

23. Interviews with Local 4 President Spohnholtz, James Armitage of Inland Press, and Arch McCready, secretary, Chicago Lithographers Association, August–September 1958.

24. Letter to the author, July 1960.

25. *Contract Discussion Transcript,* Boston, 1958, pp. 77–90.

26. *A Contract between the Contract Employers Group (Lithographers Division) and Local No. 3, Amalgamated Lithographers of America,* signed September 22, 1958, article XIV. The contract covers 15 employers. It will henceforth be cited as the *Boston Group Contract,* 1958.

27. *Agreement between the Metropolitan Lithographers Association, Inc., and Amalgamated Lithographers of America, Local No. 1,* signed June 13, 1958, article 29. This agreement will be cited as the *New York Group Contract,* 1958.

CHAPTER X. Results of Bargaining: Disrtibution of Money

1. John T. Dunlop, "The Task of Contemporary Wage Theory," *New Concepts in Wage Determination,* edited by George W. Taylor and Frank C. Pierson (New York, 1957), p. 129.

2. It is, of course, true that any survey of wage scales can be misleading for several reasons. It may ignore straight-time payments above scale, and necessarily catches some locals just coming out of negotiations and others just going in. The latter point is true of the *1947 Survey* since it was taken shortly after the New York 1946 negotiations and was in fact intended to give other locals material which would aid them in their upcoming negotiations. For this reason percentages are calculated throughout from Chicago rather than from New York.

3. This material has been provided by William Winship of Brett Litho, New York City.

4. See ALA *Proceedings,* Colorado Springs, 1946, pp. 187–194; ALA *Cleveland Policy Conference,* Cleveland, 1958, pp. 149–162.

5. Talk to Metropolitan Lithographers' Association, November 20, 1958. The quotation is from notes made at the meeting and may not be exact.

6. *Census of Manufactures: 1954,* vol. I, pp. 203–240n.

7. Letter to the author, July 1960.

8. Interview, October 1957.

9. Rates taken from relevant contracts.

10. American Can, Continental Can, National Can, Crown Cork and Seal.

11. Sutherland brief of ALA.

12. Interview with William Rheinhardt, Rand McNally, September 22, 1958.

13. ALA *Prevailing Wage Surveys, 1947* and *1956,* internal records of the union. Toledo is unique in that its lithographic industry is dominated by a trade shop, one which serves a national market.

14. H. E. Hoagland, *Collective Bargaining in the Lithographic Industry* (New York, 1917), pp. 33–34.

15. ALA *Prevailing Wage Survey,* March 1947, internal record of the union.

16. The surveys from which these figures are drawn are made up from returns by individual members, and therefore include those members who do not work in contract shops. For this reason there will nearly always be reported figures of below-scale rates.

17. For the Swedish experience, *see* Charles A. Myers, *Industrial Relations in Sweden* (Cambridge, Massachusetts, 1951), pp. 29–32. Other discussions are cited in John T. Dunlop, *Industrial Relations Systems* (New York, 1958), p. 78.

18. *New York Group Contract*, 1956, p. 28.

19. ALA *Cleveland Policy Conference*, Cleveland, 1958, p. 197.

20. Based on annual surveys conducted by the BLS, *Union Wages and Hours: Printing Industry*.

21. IPP&AU, *Convention Proceedings*, 1940, officers' reports, p. 15.

22. IPP&AU, *Convention Proceedings*, 1948, officers' reports, p. 5.

23. Material taken from *New York 1939 Group Contract*, May 31, 1939, negotiated with the Eastern Lithographers Association, predecessor of the MLA; from October 1, 1941 *Agreement* (scales only), also with the ELA; from a letter from Local 1 President Swayduck to the writer, April 14, 1959); and from the 1943 ALA *Wage Scale Survey*. The early contracts have been provided through the courtesy of Martin Grayson of Printing Developments, Inc., Chicago, and by President Swayduck.

24. Hoagland, *Collective Bargaining*, pp. 111–112.

25. NAEL general letters 1385–C (April 19, 1923), and 1386–E (June 22, 1923).

26. ALA *Proceedings*, St. Paul, 1949, officers' reports, p. 13.

27. ALA *Proceedings*, Chicago, 1957, officers' reports, p. 9.

28. Taken from March 1, 1958 booklet describing plan.

29. *Boston Group Contract*, 1958, article X.

30. The following material is taken from ALA *Proceedings*, Chicago, 1957, pp. 262–282; ALA Proceedings, Boston, 1955, pp. 242–272; *Annual Reports of the Trustees of the Inter–Local Pension Fund of the ALA*, 1958 and 1959.

31. *Minutes of Negotiations*, Boston, April 23, 1958, p. 8n.

32. ALA *Proceedings*, Boston, 1955, p. 392.

33. ALA *Proceedings*, Portland, Oregon, 1959, p. 188.

CHAPTER XI. Results of Bargaining: Distribution of Power

1. Throughout the 1920's the NAEL operated employment offices in New York, Chicago, and San Francisco which, in at least Chicago and San Francisco, were probably more effctive placement services than those the union provided.

2. Interview with John Benshop, August 1958.

3. *Agreement between* .... *and Local No. 4, Amalgamated Lithographers of America*, effective May 1, 1957, article 3. This agreement is of the type that is jointly negotiated but separately signed. The employer bargaining representative is the Chicago Lithographers Association. It will be henceforth referred to as the Chicago *Group Agreement*, 1957.

4. *Standard Working Conditions in the Lithographic Industry in Philadelphia Area under Jurisdiction of Local 14, Amalgamated Lithographers of America*, effective April 1, 1957, article II. The employers' bargaining representative is the Philadelphia Lithographers Group.

5. *New York Group Agreement*, 1958, pp. 34–35.

6. *Contract Discussion Transcript*, pp. 16–18.

7. *Agreement between Crown Cork and Seal Company, Inc. and Local No. 18 of the Amalgamated Lithographers of America* (March 1958), article II.

8. Jules Justin, "How to Preserve Management's Rights under the Labor Contract," speech at University of Michigan, Bureau of Industrial Relations conference, December 1958. Reprinted in Bureau of Industrial Relations, *Addresses in Industrial Relations* (Ann Arbor, 1959), p. 6.

9. *Minutes of Negotiations,* Boston, March 10, 1958, p. 2; March 13, 1958, p. 4; March 24, 1958, p. 8.; April 7, 1958, p. 7; June 4, 1958, language submitted by employers; June 5, 1958, p. 4.

10. ALA *Constitution,* 1958, article XIX, section 1.

11. H. E. Hoagland, *Collective Bargaining* (New York, 1917), p. 117.

12. ALA *Proceedings,* Colorado Springs, 1946, p. 98.

13. Typewritten document prepared by Local 4, Chicago, September 1946. A copy of this has been made available through the courtesy of Benjamin Robinson; Summary, p. 3.

14. ALA, *1947 National Wage Survey,* internal record of the union.

15. Internal records of the union.

16. ALA *Constitution,* article XIX.

17. *Articles of Agreement between* ———— *and Local No. 2 of the Amalgamated Lithographers of America,* effective October 1, 1956, article 19. This, like the Chicago and Boston agreements, is jointly negotiated but signed with individual firms. The group representing the employers is an informal one.

18. *New York Group Agreement,* 1958, article 18.

19. "Union President Swayduck: Profile of a New Kind of Labor Leader," *Management Methods,* November 1957, p. 97.

20. *New York Group Agreement,* 1958, p. 33.

21. ALA *Proceedings, Colorado Springs,* 1946, pp. 112–113.

22. *Boston Group Contract,* 1958, article XV; *New York Group Contract,* 1958, article 19; *Chicago Group Contract,* 1957, article 12.

23. ALA *Proceedings,* Boston, 1955, p. 199.

24. Letter from Edward Hansen to Robert Bruck, March 1, 1946.

25. ALA *1947 Special Negotiating Session.*

26. *Chicago Group Contract,* 1953.

27. National Association of Photo-Lithographers, *Wage Scales, Working Conditions and Complements of Press Help Covering the Lithographic Industry.* Revised to October 25, 1957. This summary is a regular service to members of the NAPL, and has been made available through the courtesy of Walter Soderstrom, the executive vice-president. *See also* ALA *Proceedings,* Chicago, 1957, officers' reports, p. 9.

28. BLS, *Union Wages and Hours: Printing Industry,* bulletin no. 1228 (May 1958), pp. 15–17.

29. *See* preceding table.

30. Interview, November 1958.

31. ALA *Proceedings,* Portland, Oregon, 1959, p. 325.

32. Mimeographed release of BSL. Data available from *Monthly Labor Review* each month, statistical series C-1.

33. Average weekly hours take into consideration short-time, absenteeism, and holidays not worked, as well as overtime.

34. *Boston Group Contract,* 1958, article VII.

35. *Contract Discussion Transcript,* pp. 46–47.

36. *Boston Group Contract,* 1958, article VII.

37. *Minutes of Negotiations,* Boston, February 27, 1958, p. 7; article ix; article xvi.

38. ALA *Proceedings,* Portland, Oregon, 1959, pp. 326–327.

39. Benjamin Robinson, *Memorandum to International Council,* March 6, 1944.

40. *Minutes of Negotiations,* Boston, March 6, 1958; April 2, 1958, p. 4.

41. The following sources have been used for the San Francisco strike and court actions: *Printing Impressions,* 2.9–11 (February, March, and April 1960); *Lithographers' Journal,* January and February 1960; interview with ALA President Kenneth Brown, March 1960; U.S. District Court, Brown *v.* Local 17, Lithographers (January 13, 1960), reported in Bureau of National Affairs *Labor Relations Reference Manual,* 45 LRRM 2578.

42. Gerald Reilly, now with NAPL, reported in *Printing Impressions* that the ALA had lost the fight, an opinion which Benjamin Robinson challenged in the next month's issue. For citation see preceding note.

CHAPTER XII. The System: Present and Future

1. NAEL general letter 1386-F (June 26, 1923).

2. From notes of a talk to the Metropolitan Lithographers Association (MLA), November 20, 1958. Quote is from notes taken at the meeting, and may not be exact.

3. The article appears in E. Wight Bakke *et al., Labor Mobility and Economic Opportunity* (New York, 1954), pp. 93–109.

4. Since writing this I have learned that precisely such a service is being rendered in certain labor markets by the California Department of Employment in cooperation with the affected trade unions and employers. The program is described in the *Employment Security Review,* 28.11:24–26 (November 1961).

Index

Accounting, 14–15, 17, 58–59
Ace Folding Box Co., 178
Advertising industry: integration of lithography with, 7, 63–64
ALA, *see* Lithographers of America, Amalgamated
Allied Printing Trades Assn., 143
Amalgamation, *see* Merger
American Federation of Labor (AFL): in offset press controversy, 98; awards jurisdiction to printing pressmen, 132, 133; left by ALA, 134
American Federation of Labor Congress of Industrial Organizations (AFL-CIO): left by ALA, 71; position on bargaining units, 146
American Lithographic Co., 6
American Potash case, 135
Antiunion activities, 92–94, 166
Apprenticeship: cooperative programs, 164; duties in, 200; nationwide bargaining on, 202–203; regulation of, 76, 107, 117, 137, 163, 198–203; training, 76
Arbitration, 88, 91, 169–170
Argus Publishing Co., 127
Armitage, James, 161
Artists and Engravers Assn., 86
Artists', Engravers', and Designers' League, 88
Automobile Workers, United, 25

Baker, Elizabeth, 143
"The Balkinization of Labor Markets," 230
Baltimore Can Co., 194
Benefit Plans, 170: union administration of, 189–191
Berry, George, 98
Bien, Julius, 87
Blackburn, John, 146, 148, 154, 167
Bock, Philip, 103, 139
Bonuses, 105
Boycotts, secondary, 165, 215
Brandenberg, Ivan T., 184
Brandt, Theodore, 184
Brokers, printing, 68–69
Brown, Kenneth, 148
Bruck, Robert, 121, 146, 148, 226
Business depression: 1920, 102; 1930–1933, 113
Buyer-supplier bond, 63, 64–66, 168; effect

on industry organization, 66–68; in commercial printing, 71

Canadian Lithographers Assn., 18–19, 82
Canary, George, 28–30, 141–142, 147, 148, 154, 161
Captive plants, 9–11
Castro, Albert, 114, 121
Central Lithographic Trades Council, 89, 90; degeneration of, 92
"Chain shop" clause, 215, 216
Check-off, 214
Clayton Act, 3
Collective bargaining: annual, 119; "break-off point," 162; in Canada, 18–19; future course of, 223–224; government role in, 150; impact of NRA, 120–121; informal, 161, 163; local level, 152, 219; national, 149–150, 152–153, 223; "one man," 162; procedures, 34–35; temper of, 104, 159–160
    employer associations in: 17, 19–25; coordination, 53–54, 70, 160
    issues: 35, 70, 157, 159–161; 1919, 100–101; 1921, 103; 1946, 137; 1950's, 139; World War II, 128; Canada, 1948, 138
    strategy: "cushions," 162; employer, 105, 153–154, 165–166; union, 130, 154–159, 164–165
    unions in: attitude toward, 1917, 100; coordination, 103–104, 158–161; policies, 24; power, 180
Collective bargaining unit, 131, 132, 180, 214, 222, 223; "carve out," 135, 145–146; "traditional," 133, 135, 151, 221
Collective labor agreements: administration, 168–170; legal status, 165, 215–217; national, 1, 17–19, 90–91, 100–102, 121, 122; Canada, 18–19; supplemental, 122
Color printing, 54
Colwell Press, 127
Combination plants, 12–13, 17, 19, 48, 126; workforce in, 72
Commercial printing: lithography in, 54
Competition: 48, 52, 179, 209, 214; bidding, 59–60; control of, 117; nonprice, 62; price, 87
Congress of Industrial Organizations (CIO): joined by ALA, 134

WERTHEIM PUBLICATIONS
IN INDUSTRIAL RELATIONS

PUBLISHED BY HARVARD UNIVERSITY PRESS

J. D. Houser, *What the Employer Thinks*, 1927
Wertheim Lectures on Industrial Relations, 1929
William Haber, *Industrial Relations in the Building Industry*, 1930
Johnson O'Connor, *Psychometrics*, 1934
Paul H. Norgren, *The Swedish Collective Bargaining System*, 1941
Leo C. Brown, S.J., *Union Policies in the Leather Industry*, 1947
Walter Galenson, *Labor in Norway*, 1949
Dorothea de Schweinitz, *Labor and Management in a Common Enterprise*, 1949
Ralph Altman, *Availability for Work: A Study in Unemployment Compensation*, 1950
John T. Dunlop and Arthur D. Hill, *The Wage Adjustment Board: Wartime Stabilization in the Building and Construction Industry*, 1950
Walter Galenson, *The Danish System of Labor Relations: A Study in Industrial Peace*, 1952
Lloyd H. Fisher, *The Harvest Labor Market in California*, 1953
Theodore V. Purcell, S.J., *The Worker Speaks His Mind on Company and Union*, 1953
Donald J. White, *The New England Fishing Industry*, 1954
Val R. Lorwin, *The French Labor Movement*, 1954
Philip Taft, *The Structure and Government of Labor Unions*, 1954
George B. Baldwin, *Beyond Nationalization: The Labor Problems of British Coal*, 1955
Kenneth F. Walker, *Industrial Relations in Australia*, 1956
Charles A. Myers, *Labor Problems in the Industrialization of India*, 1958
Herbert J. Spiro, *The Politics of German Codetermination*, 1958
Mark W. Leiserson, *Wages and Economic Control in Norway, 1945–1957*, 1959
J. Pen, *The Wage Rate under Collective Bargaining*, 1959
Jack Stieber, *The Steel Industry Wage Structure*, 1959
Theodore V. Purcell, S.J., *Blue Collar Man: Patterns of Dual Allegiance in Industry*, 1960
Carl Erik Knoellinger, *Labor in Finland*, 1960
Sumner H. Slichter, *Potentials of the American Economy: Selected Essays* edited by John T. Dunlop, 1961
C. L. Christenson, *Economic Redevelopment in Bituminous Coal: The Special Case of Technological Advance in United States Coal Mines, 1930–1960*, 1962
Daniel L. Horowitz, *The Italian Labor Movement*, 1963

Studies in Labor-Management History

Lloyd Ulman, *The Rise of the National Trade Union: The Development and Significance of its Structure, Governing Institutions, and Economic Policies*, 1955
Joseph P. Goldberg, *The Maritime Story: A Study in Labor-Management Relations*, 1957, 1958
Walter Galenson, *The CIO Challenge to the AFL: A History of the American Labor Movement, 1935–1941*, 1960